MICROSENSORS

MICROSENSORS
PRINCIPLES AND APPLICATIONS

Julian W. Gardner
University of Warwick, UK

JOHN WILEY & SONS
Chichester · New York · Brisbane · Toronto · Singapore

Copyright © 1994 by John Wiley & Sons Ltd
Baffins Lane, Chichester
West Sussex PO19 1UD, England
National Chichester (0243) 779777
International (+44) 243 779777

Reprinted December 1994

Other Wiley Editorial Offices

John Wiley & Sons, Inc., 605 Third Avenue,
New York, NY 10158-0012, USA

Jacaranda Wiley Ltd, 33 Park Road, Milton,
Queensland 4064, Australia

John Wiley & Sons (Canada) Ltd, 22 Worcester Road,
Rexdale, Ontario M9W 1L1, Canada

John Wiley & Sons (SEA) Pte Ltd, 37 Jalan Pemimpin #05-04,
Block B, Union Industrial Building, Singapore 2057

Library of Congress Cataloging-in-Publication Data

Gardner, J. W. (Julian W.), 1958–
    Microsensors : principles and applications / J. W. Gardner.
        p. cm.
    Includes bibliographical references and index.
    ISBN 0 471 94135 2 — ISBN 0 471 94136 0 (pbk.)
    1. Transducers. 2. Detectors. 3. Microelectronics. I. Title.
TK7872. T6G37 1994
    681.2—dc20   94-10066
                                                           CIP

British Library Cataloguing in Publication Data

A catalogue record for this book is available from the British Library

ISBN 0 471 94135 2; 0 471 94136 0 (pbk.)

Produced from camera-ready copy supplied by the author using Word for Windows
Printed and bound in Great Britain by Bookcraft (Bath) Ltd.

Contents

Preface

The aims of this book are to discuss the fundamental principles of miniature transducers (microsensors), to provide an account of recent developments and to illustrate, with examples, their application.

In the past, the name *transducer* was used for devices that either converted a non-electrical physical or chemical quantity, e.g. pressure, temperature, humidity, into an electrical signal <u>or</u> vice versa.[1] However, it is now the custom to call the type of transducer that converts a non-electrical quantity into an electrical signal a *sensor*, and the type of transducer that converts an electrical signal into a non-electrical quantity an *actuator*. The electrical signal coming from the sensor often needs modification before it can play a useful role, such as displaying information to the operator or initiating a control process. Any device that modifies the electrical signal from the sensor, e.g. an operational amplifier or an analogue-to-digital converter, may be called a *processor* (although it is occasionally referred to as an electric transducer). Recent improvements in both our understanding of the sensing principles and the enabling technologies have led to the design of superior sensors with, for instance, linear (analogue) outputs that can be readily utilised. Thus, sensors, actuators and processors are now used in a wide range of measuring instruments that have a variety of uses from the home to the factory floor.

A reduction in the size of a sensor can often lead to an increase in its applicability and a lower cost of manufacture may be achieved due to a wider marketability. Moreover, the silicon revolution has enabled us to produce small, low cost, reliable processors in the form of Integrated Circuits (ICs) using microelectronics technology. This recent advance has led naturally to a considerable demand for small sensors or *microsensors* that can fully exploit the benefits of the IC processor technology. These benefits include the ability to manufacture at low cost large numbers of reliable but sophisticated microsensors. For the purpose of writing this book, I use the term "microsensor" to mean a sensor that has a physical dimension at the sub-millimetre level and not a sensor that measures very small quantities. However, it often turns out that the dynamic range of such microsensors far exceeds that of larger sensors. The principal microsensors described here are made using conventional thin and thick film

---

[1] See Dictionary of Electronics (ed. V. Pitt), Penguin Books Ltd, Middlesex, UK.

technologies as well as more recent techniques, e.g. silicon micromachining. The latter technique is particularly promising because it allows the full integration of the sensor, processor and even actuator on a single silicon wafer.

Thus, following the introduction and a chapter on basic signal processing, this book focuses on microsensors - the processing of materials for their fabrication, the consideration of the wide range of physical, chemical, mechanical principles employed, and the benefits incurred from the use of *smart sensors* and *microsensor array devices.*

The book has been written as a text suitable for final year undergraduate courses and for postgraduate students.[2] In addition, it is also intended to provide valuable background for scientists and engineers who currently employ conventional transducer technology but would benefit significantly from the use of microsensor technology. Further reading is suggested at the end of each chapter, together with a full list of references and a set of problems.

Several appendices may be found at the back of this book which may be of help to the reader. These provide lists of symbols, selected definitions and acronyms, base SI units, unit prefixes for SI units, a table of Laplace transforms, unit conversion factors and the values of fundamental constants.

Acknowledgements

I wish to thank many of my colleagues for their help in preparing this book. In particular, I wish to thank Dr Peter Jones, Dr Neil Storey, Professor Keith Bowen, Dr Sam Davies, Dr David Hutchins, Dr Derek Chetwynd, Dr David Gordon-Smith, Professor Philip Bartlett and Dr Neil White for proof reading the relevant chapter. I also wish to thank the group secretary, Gill Pearce, who patiently typed out the manuscript. Finally, I am deeply indebted to Dr William Gardner for his generous help and advice in preparing this book.

Julian William Gardner
University of Warwick

---

[2] Suitable for Level 5 courses in the USA.

1. Introduction

Objectives

☐ To introduce the basic concepts in measurement systems

☐ To define sensor terminology

☐ To identify sensor applications

☐ To present the need for microsensors

1.1 Measurement Systems

The human appetite for information is almost limitless: an appetite that is nourished after birth and rapidly grows through infancy to adulthood. However due to the enormous amount of information man needs to be a sophisticated gatherer and in order to make use of it, man also needs to be an efficient processor of information. Man gathers information from the surroundings using an assortment of measurement systems that are called "senses". The traditional senses are sight, hearing, touch, smell and taste which are used to measure physical and chemical quantities in the environment. These measurements are then processed by a massively parallel neural architecture (the brain) and used to make decisions, e.g. which food to eat. Man's biological measurement system permits the completion of numerous simple and complex tasks in everyday life. It has remarkable capabilities, such as its ability to process rapidly a vast amount of data in an adaptive fault-tolerant manner. Yet, the information is by its very nature subjective and qualitative. So man has historically obtained advantages from the design and use of instruments and tools which provide a quantitative measurement system. Measurement systems have been designed which employ physical or chemical properties in order to obtain quantitative data which can then be processed in one of several ways.

1

Figure 1.1 shows the basic functions of a measurement or information-processing system. The input signal to a measurement system is often called the *measurand* which is the physical or chemical quantity to be measured (e.g. displacement, pressure, gas concentration). The measurand is detected or sensed by what is normally called a *sensor* or *input transducer.* A sensor may be defined as a device that converts a non-electrical physical or chemical quantity into an electrical signal. The electrical signal from the sensor often needs modification before it can serve a useful function, such as displaying information to an operator, being recorded on some medium or transmitted somewhere else. Most signals coming from a sensing element are analogue in nature, and so analogue signal processing is generally needed. A *processor* may be defined as any device that modifies the electrical signal coming from the sensor - without changing the form of the energy that describes the signal. Often it is useful to distinguish between the main processing unit (e.g. an Intel 486 microprocessor) and a signal preparation unit (e.g. an amplifier, filter or analogue to digital converter). I shall refer to this type of device as a *preprocessor* or *converter.* Both preprocessors and processors can play an important role in a measurement system. Some of these devices are discussed in Chapter 2 together with a description of some common interface systems.

Finally, the signal from the processor is used to display some information to an operator, e.g. drive a Liquid Crystal Display (LCD) Unit or Visual Display Unit (VDU). Alternatively, the signal might be recorded, for example, on paper, magnetic disc or an IC chip (RAM, EPROM, etc.). Any such device that converts an electrical signal into a physical or chemical quantity will be referred to here as an *actuator* or *output transducer.* For example, the display of information on a VDU requires the conversion of an electrical signal into an optical signal and is thus an actuating process. The optical signal may then be detected, modified and acted

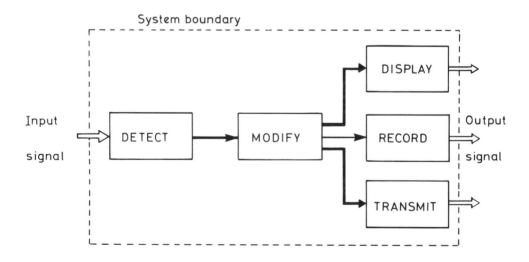

Figure 1.1 Functional block diagram of a measurement system.

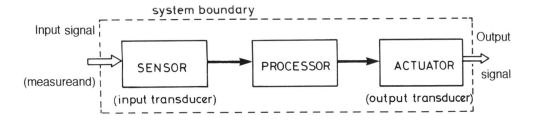

Figure 1.2 Basic components of a measurement or information-processing system.

upon by a human operator.

Clearly, there can be many stages to an information-processing system but Figure 1.2 shows its most basic components: a sensor, a signal processor (including the preprocessor) and an actuator. This book is primarily concerned with sensors, although some discussion of the other two basic components is made when required.

1.2 Classification of Sensing Devices

It is necessary when discussing the subject of sensors, to decide whether to classify them according to the function that they perform (e.g. the measurement of pressure, temperature etc.) or the physical principle upon which they work (e.g. magnetoresistive, optoelectronic etc.). It is now common practice to classify sensors in terms of the main forms of energy that carry the signal. Table 1.1 shows the various forms of energy that are normally used to categorise sensors, together with some of their typical measurands.

The layout of this book has been designed to reflect the classification of sensors into their principal forms of signal: thermal, radiation, mechanical, magnetic, chemical, biological and electrical. The last class of sensor, electrical, is treated here as an electric transducer or processor, rather than as a sensor because there is no conversion in the form of energy of the signal. This leads to the individual chapters being ordered according to the function that the sensors perform (for example, thermal sensors for measuring temperature or mechanical sensors for measuring pressure). The advantage of classifying sensors by their function rather than by their principle of operation is that normally the reader wishes to learn about the different types of pressure sensors available instead of their underlying principles. I believe this to be the easiest and most practical layout to adopt.

As mentioned earlier, the biological measurement system of man has the five principal senses of sight, hearing, touch, smell and taste. These five senses can be

Table 1.1 Classification of sensors by signal form.

Form of signal	Measurands
Thermal	Temperature, heat, heat flow, entropy, heat capacity etc.
Radiation	Gamma rays, X-rays, ultra-violet, visible, infra-red, micro-waves, radio waves, etc.
Mechanical	Displacement, velocity, acceleration, force, torque, pressure, mass, flow, acoustic wavelength and amplitude etc.
Magnetic	Magnetic field, flux, magnetic moment, magnetisation, magnetic permeability etc.
Chemical	Humidity, pH level and ions, concentration of gases, vapours and odours, toxic and flammable materials, pollutants etc.
Biological	Sugars, proteins, hormones, antigens etc.
Electrical[1]	Charge, current, voltage, resistance, conductance, capacitance, inductance, dielectric permittivity, polarisation, frequency etc.

[1] "Electrical sensors" are classified here as electric transducers or processors rather than sensors, as there is no conversion of energy form of the signal.

described according to the signal form in which the energy is carried, the measurand and the sensing device. Table 1.2 summarises these classifications and lists some analogous devices to the biological sensors. For example, the role of the retina in the human eye may be mimicked by devices such as photodiode arrays (§12.2.2).

This classification system can also be used to categorise the types of actuators that we use in a measurement system. For example, Table 1.3 shows how the actuator functions given in Figure 1.1 of display, record and transmit can be classified according to the form of their signals and the nature of their generation.

Table 1.2 Classification of the human senses.

Human sense	Signal	Measurand	Sensing device	Analogue device
Sight	Radiant	Intensity and wavelength of light.	Rods and cones in retina.	Photographic film, photodiode, phototransistor.
Hearing	Mechanical	Intensity and frequency of sound.	Cochlea in inner ear.	Microphone.
Touch	Mechanical	Pressure, force.	Nerves.	Potentiometers and LVDTs (simple touch), optical gauging and tactile arrays (complex touch).
Smell	Chemical	Odorants.	Olfactory receptor cells in nose.	Electronic nose.
Taste	Biological	Proteins.	Taste buds in tongue.	

Table 1.3 Classification of some common actuators.

Function	Actuator	Signal	Principle
Display	Light emitting diode.	Radiant	Current generation of photons.
	Visual display unit.	Radiant	Fluorescent screen.
	Liquid crystal display.	Radiant	Transmittance of polarised molecular crystals.
Record	Thermal printer.	Thermal	Ink is melted.
	Magnetic recording head.	Magnetic	Magnetisation of thin films on computer disc.
	Laser.	Radiant	Ablation of material on optical disc.
Transmit	Loudspeaker.	Mechanical	Generation of sound.
	Aerial.	Radiant	Generation of radio waves.
	Electric motor.	Mechanical	Generation of motion.

Thus the function of "display" can be a Light Emitting Diode (LED) that is illuminated to show the presence of an object or occurrence of a fault.

Some sensors are said to be *self-generating* or *self-exciting* as opposed to *modulating*. A self-generating sensor is one which does not need an external power supply to work. For example, a thermocouple (§5.2.2) is a self-generating thermal sensor because it produces an e.m.f. from the difference in junction temperatures. In this case, energy is supplied by the thermodynamic system rather than any external power supply. In contrast, a photodiode (§6.3.3) is a radiant sensor whose forward bias current is modulated by photo-induced electrons. Self-generating sensors tend to produce very low output powers and so often need (pre)processors to amplify the signal strength to a useful level, whereas modulating sensors (e.g. photodiodes, photocells or thermistors) with a higher efficiency tend to produce much higher output energies. However, it is still usual to further process the output signals from modulating sensors to obtain standard electrical signals (e.g. a current range of 4 to 20 mA, or a voltage range of 0 to 100 mV, see §2.3).

1.3 Ideal Sensor Characteristics and Practical Limitations

A sensor in its simplest form may be regarded as a system with an input $x(t)$ and output $y(t)$. Figure 1.3 shows a system representation of (a) a self-exciting sensor and (b) a modulating sensor. A self-exciting sensor has its output energy supplied entirely by the input signal $x(t)$ and good examples of this are a thermocouple (§5.2.2) and a piezoelectric crystal (§7.7.2).

For a thermocouple, the input signal or measurand is the difference in junction temperatures $\Delta T(t)$ and the output is the e.m.f. $\phi(t)$ (in volts).

The general equation that describes a self-exciting sensor system is

$$y(t) = F(x(t)) \tag{1.1}$$

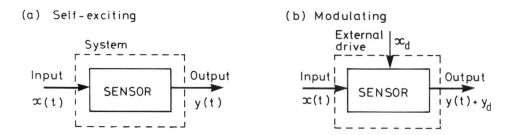

Figure 1.3 Basic representation of self-exciting and modulating sensor systems.

or in the case of a thermocouple, more specifically

$$\phi(t) = F(\Delta T(t)) \tag{1.2}$$

where $F(x(t))$ is the characteristic relationship that describes the behaviour of a self-exciting sensor. In the case of the modulating sensor, the system equation (1.1) can be written more explicitly as

$$y(t) = F\left(x(t) + x_d\right) \tag{1.3}$$

where the external supply signal $x_d(t)$ should ideally be stationary and noise free. For example, $x_d(t)$ could be a constant driving current or applied reference voltage.

Figure 1.4 shows the ideal input-output characteristic of a sensor in which the input signal (measurand) is directly proportional to the output signal. The ideal sensor not only has a linear output signal $y(t)$ but it should instantaneously follow the input signal $x(t)$, whence

$$y(t) = S.x(t) \tag{1.4}$$

The slope S of the input-output curve has a constant value for a linear sensor and is usually referred to as the *sensitivity*. In practice, equation (1.4) must be used carefully as no sensor can respond instantaneously to a change in the input signal but requires some time to reach its steady-state value, e.g. the system needs time to find a new thermal, electrical or chemical equilibrium. However, if we assume that we have a time-dependent *linear* sensor system, then we can re-write equation (1.1) more explicitly by,

$$a_n \frac{d^n y}{d t^n} + a_{n-1} \frac{d^{n-1} y}{dt^{n-1}} + \ldots + a_1 \frac{dy}{dt} + a_0 y = x(t) \qquad (1.5)$$

where a_i are the i linear coefficients which can be physically interpreted. We can now see that a self-exciting sensor has zero output for zero input (i.e. $y(t)=0$ when $x(t)=0 \ \forall \ t$), whereas a modulating sensor has a non-zero output of $y_d(t)$ when the input is zero (i.e. $y(t)=y_d(t)$ when $x(t)=0 \ \forall \ t$).

Now often a simple sensor, such as a thermistor, may be modelled as a first order linear sensor with all linear coefficients higher than a_1 set to zero. The physical reason for this is that a thermal system is almost exclusively first order because there is only one energy-storing element (heat capacity which is assumed temperature independent). Moreover, we have assumed that the heat loss is linear (which is not strictly true). Consequently, the ideal characteristic of such a sensor, again in general notation, is given by

$$a_0 y + a_1 \frac{dy}{dt} = x(t) \qquad (1.6)$$

where the coefficient a_0 relates to the system gain and a_1 relates to its characteristic time response. Equation (1.6) may be solved by the method of Laplace transforms for any linear input, $x(t)$. In the case of an instantaneous change in the measurand from zero to x_m, i.e. a step change in the input signal we have,

Input: $x(t) = x_m \mu(t)$ \qquad (1.7)

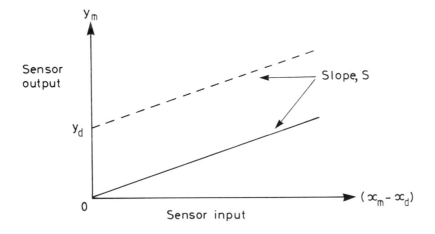

Figure 1.4 Ideal input-output relationship of self-exciting (solid-line, $x_d=0$) and modulating (dashed line, $x_d \neq 0$) sensors.

where $\mu(t)$ is the unit step function at $t=0$ and x_m is the height of the step. Taking equation (1.6) to describe a first order system, we can solve for the response of a linear sensor by taking the Laplace transform of both sides to give,

$$a_0 Y(s) + a_1 [sY(s) - y(+0)] = X(s) \qquad (1.8)$$

where the initial value of the output $y(+0)$ is taken here as zero for a self-exciting sensor, but could be rewritten as y_d for a modulating sensor.

The transfer function $H(s)$ of a self-exciting (or other) system can be written in general as,

$$H(s) = \frac{Y(s)}{X(s)} \qquad (1.9)$$

and so the transfer function of our first order linear sensor system is given by

$$H(s) = \frac{1}{(a_0 + a_1 s)} = \frac{1/a_0}{1 + (a_1/a_0)s} \qquad (1.10)$$

Thus the ratio of the coefficients (a_1/a_0) is the characteristic response time τ of the sensor and the gain is $1/a_0$. So the output of the sensor $Y(s)$ in Laplace space to a step input of height x_m is,

$$Y(s) = H(s)X(s) = \frac{(x_m/a_0)}{s(1 + \tau s)} \qquad (1.11)$$

Taking the inverse Laplace transform (from Appendix E) gives,

$$y(t) = \frac{x_m}{a_0}[1 - \exp(-t/\tau)] \qquad (1.12)$$

Thus, the behaviour of a self-exciting linear sensor is analogous to that of an electrical *RC* circuit or mechanical mass-damper network. Similarly, the ideal dynamic behaviour of a modulating linear sensor is

$$y(t) = \frac{x_m}{a_0}[1 - \exp(-t/\tau)] + y_d \qquad (1.13)$$

where y_d is the output signal generated by an external DC supply.

There are several important sensor parameters that we can now define and illustrate using this ideal sensor system. For example, the *response* Δy of a sensor is the change in output signal and is given by

$$\Delta y(t) = (y(t) - y_d) = \frac{x_m}{a_0}[1 - \exp(-t/\tau)] \qquad (1.14)$$

for both self-exciting and modulating linear sensors. In addition, the *gain* A (and sensitivity *S*) of any sensor can be defined as the absolute value of (and change in) output signal relative to the absolute value of (and change in) input signal, whence

$$A = \frac{y}{x} \quad \text{and} \quad S = \frac{dy}{dx} \tag{1.15}$$

So the sensitivity of an ideal sensor is equal to the gain and independent of both the amplitude x_m and the angular frequency ω of an AC input signal; this is also true in a real sensor provided the input frequency is below the band-width of the sensor (i.e. $\omega < 1/\tau$). But here the response of this type of sensor to any input signal $y(t)$ may be calculated from the transfer function $1/(1+\tau s)$ and is in essence a first order low-pass *RC* filter. This model is a particularly useful approximate description of a sensor that can be used within a complex control system.

Figure 1.5 shows a sketch of our calculated response of an idealised linear sensor to a step increase in the measurand signal $x_m \mu(t)$. From this sketch and Figure 1.4, we can readily see the desirable (i.e. ideal) characteristics of sensors and we will then introduce some of the undesired characteristics that often occur in everyday sensors.

As mentioned previously, the response Δy of a sensor is the change in output signal, i.e. $(y_m - y_0)$, using the notation of y_0 to represent the output of the unperturbed sensor at $x=0$, rather than the less general term of $y_d(t)$. Then y_0 represents the *baseline* signal, and should ideally be independent of time or any other physical property of the system. The baseline can be regarded as the output under certain stated conditions, e.g. a standard temperature or pressure, and need not be the zero point for a self-exciting sensor. For example, the baseline could be the e.m.f. generated by a thermocouple at room temperature. Often the baseline signal is effectively a systematic *offset* voltage or current and is removed by placing the sensor in a bridge circuit (§2.2.1).

The rise (and fall) of an output signal from a sensor is ideally exponential with a characteristic time, τ, of (a_1/a_0), see equation (1.13) with $y_d=y_0$. The characteristic time constant τ can usually be related to the physical properties of the system (e.g. the ratio of the heat capacity and thermal diffusivity in a first order thermal system). Often we define the time taken for the sensor signal to reach 90% of its final value and this is referred to as the t_{90} time or sometimes the response time. It is desirable for the value of the t_{90} time to be less than a few seconds so that the sensor reaches a final reading in a practical time.

The sensitivities of both self-exciting S_s and modulating S_m sensors are by definition from equation (1.15) the same, i.e.

$$S_s = S_m = \frac{dy}{dx} \tag{1.16}$$

In the case of a linear first order sensor the sensitivity is the steady state response, $(y_m - y_0)$ divided by the input $(x_m - x_0)$ and is equal to the gain, but in non-linear sensors the two values differ. In principle such a sensor would have an

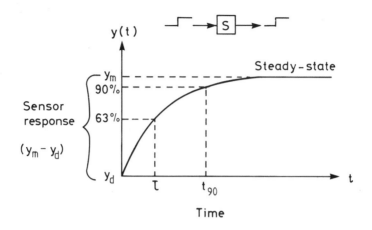

Figure 1.5 Characteristic or transient response of an ideal (i.e. linear, first order) sensor system.

infinite range of possible measurements, i.e. it could measure all values of the measurand from zero to infinity. In practice, the working *range* of a sensor is limited and the sensitivity falls off above some characteristic value, see Figure 1.6.

For example, the working range may be limited by the associated electronic circuitry as the voltage output from an op-amp approaches that of its supply rail. From the input-sensitivity curve, one can see the working range $(x_{max} - x_{min})$ of the sensor and thus design the desired full-scale reading y_{max} (e.g. a current of 20 mA) near the limit of its working range x_{max}.

Table 1.4 summarises the desirable characteristic features of a sensor mentioned above and gives their ideal values.

Table 1.4 Desirable sensor characteristics.

Characteristics	Ideal value
Response, Δy	Exactly linear and noise free.
Baseline, y_0	Zero point.
Response time, τ	Zero for instantaneous response.
Frequency band-width	Infinite for instantaneous response.
Time to reach 90% of final value, t_{90}	Zero for instantaneous response.
Full-scale reading, y_{max}	Calibrated max. output, e.g. 20 mA current.
Working range, $(y_{max} - y_{min})$	Infinite.
Sensitivity, S	High and constant over entire working range.
Resolution	Infinite.

When considering real sensors, rather than idealised ones, we find that there are some common but undesirable characteristics to consider. These are summarised in Table 1.5 below and now briefly discussed.

Table 1.5 Undesirable sensor characteristics.

Characteristic	Meaning
Non-linearity	Response is not proportional to the input signal.
Slow response	Output is slow to reach a steady-state value (i.e. large value of τ).
Small working range	Operating range is highly restricted.
Low sensitivity	Sensor can only respond to large input signals (i.e. S is small).
Sensitivity drift	Output varies with time, e.g. change in ambient temperature.
Baseline drift	Output varies with time.
Offset	Systematic error in sensor output.
Offset drift	Offset drifts in time, e.g. due to sensor ageing.
Ageing	Sensor output changes with time.
Interference	Output is sensitive to external conditions, e.g. stray electromagnetic radiation, humidity etc.
Hysteresis	Systematic error in the input-output curve.
Noise	Output contains an unwanted random signal.

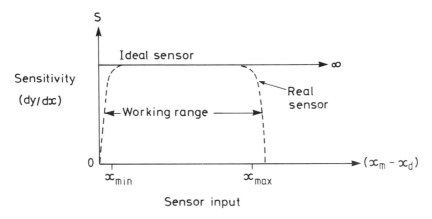

Figure 1.6 Input-Sensitivity relationship for ideal (solid-line) and real (dashed-line) sensor systems.

1.4 Sensors in Product and Process Control

Clearly, sensors form an essential part of a measurement or information-processing system (§1.1). Moreover, we can also use the output from a sensor to drive a process and, in effect, control it.

Let us suppose that we have obtained a voltage signal $v(t)$ from a sensor that we wish to use as a command variable to control the position of a motor. Figure 1.7 shows the elements of a DC-motor (i.e. the actuator) in which the input voltage $v(t)$ is amplified and used to drive the motor armature producing a torque τ_m. The

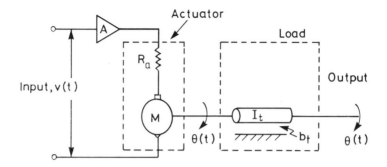

Figure 1.7 Linear model of a DC motor and load system.

motor drives a load which is rotated by an angle $\theta(t)$. Thus the response of the system is the angle $\theta(t)$ through which the load moves. Figure 1.8 (a) and (b) show a schematic of the open-loop control of an actuator together with a description of the process using standard control nomenclature. In open-loop control, the output signal $Y(s)$ is related to the input signal $V(s)$ by,

$$Y(s) = G_p(s)\, G_m(s)\, V(s) \tag{1.17}$$

where $G_p(s)$ and $G_m(s)$ are the transfer functions of the controller/amplifier (i.e. processor) and the motor/load elements, respectively, and $V(s)$ is the Laplace transform of the input voltage signal $v(t)$. The transfer function for the system $G(s)$ is given by $Y(s)/V(s)$, i.e. $G_p(s)\, G_m(s)$.

For the idealised model of a DC motor given in Figure 1.7, the transfer function may be written as

$$\frac{Y(s)}{V(s)} = \frac{A}{s(1+\tau s)} \tag{1.18}$$

where A is the static gain of the system with

$$A = \frac{k_m}{(b_t R_a + R_m k_b)} \tag{1.19}$$

where R_a is the armature resistance, R_m the torque constant, k_b the back e.m.f. coefficient and b_t the total viscous friction. The time constant of this circuit τ is related to the total moment of inertia on the motor shaft I_t and friction coefficients:

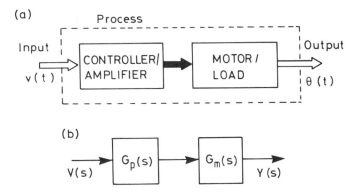

Figure 1.8 (a) Schematic representation of an open-loop DC motor with (b) its control block diagram.

$$\tau = \frac{I_t}{(b_t + k_m k_b / R_a)} \qquad (1.20)$$

The problem with open-loop control of a process variable is that there is no way of knowing that the output $\theta(t)$ is at the desired value. The performance of open-loop control thus depends upon the stability of the controller/amplifier and motor/load elements. Systematic errors can occur in such a system due to ageing of components, temperature drift, etc. Consequently, closed-loop control is often preferable in manufacturing or processing products. Closed-loop process control requires the use of a sensing device in the feedback line as shown in Figure 1.9(a). From the control circuit in Figure 1.9(b), we can calculate the difference $e(s)$ between the output demanded to that measured by the sensor, where

$$e(s) = V(s) - H(s)Y(s) \qquad (1.21)$$

Under steady-state conditions there should be negligible error (i.e. $e(s) \approx 0$ as $t \rightarrow \infty$) and the transfer function of the system now includes the dynamic properties of the sensor $H(s)$, where,

$$\frac{Y(s)}{V(s)} = \frac{G(s)}{1 + G(s) H(s)} \qquad (1.22)$$

where $G(s)$ is $G_p(s)G_m(s)$. Clearly, it is important that we understand the dynamic response of a sensor in order to utilise them in controlling plant or processes in industry. The transfer function of a first order linear sensor $H_1(s)$ (from equation (1.10)) is given by

$$H_1(s) = \frac{A_s}{(1 + \tau_s s)} \qquad (1.23)$$

(a)   (b)

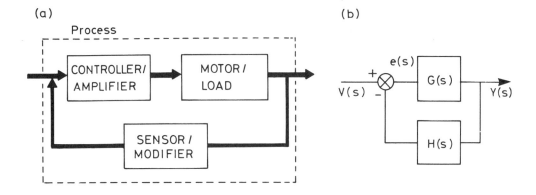

Figure 1.9 (a) The use of a position sensor to obtain closed-loop control of a DC motor system, and (b) the control block diagram with a motor element $G(s)$ ($\equiv G_p(s).G_m(s)$) and feedback element $H(s)$.

where A_s represents the steady-state gain and τ_s the time constant. So provided that $\tau_s < \tau$, then the response of the overall system remains constant.

Simple models are very useful and can give approximate solutions to sensor problems in automated manufacturing and process industries. Nevertheless there are cases in which a more sophisticated approach is needed using non-linear and/or adaptive control systems. However, such approaches are outside the scope of this book.

1.5 Why Microsensors?

Clearly, sensors have an important role to play in our everyday lives in which we have a need to gather information, process it and perform some task. Yet, the successful application of a sensor is determined by its performance, cost and reliability.[1] Figure 1.10 shows the considerable success already achieved in the UK market in 1990 and the projected growth by the year 2000 [1.1]. The UK market is subdivided into thirteen major application areas with plant and process being the largest worth £22 million in 1990 and probably £46 million in 2000.

Nevertheless a large sensor, such as a photomultiplier tube, may have excellent operating characteristics but its marketability is severely limited simply by its size. A reduction in the size of a sensor often leads to an increase in its applicability through:

- lower weight (greater portability);
- lower manufacturing cost (less materials);
- wider range of applications.

---

[1] Reliability is defined as the ability of an item to perform a required function under stated conditions for a stated period of time (see reference [1.2] and §10.3).

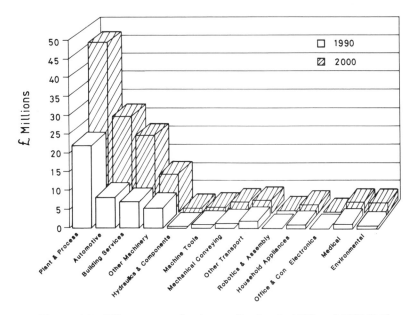

Figure 1.10 UK sensor market by application in 1990 and 2000 [1.1].

The cost of a sensor is often the single most important factor. Clearly, less materials are needed to manufacture a small sensor but the cost of materials' processing is often a more significant factor. The silicon revolution has enabled us to produce small, reliable processors in the form of Integrated Circuits (ICs) using microelectronic technology. Silicon processing technology (Chapter 3) is arguably the most advanced technology in the world, having been established about 30 years ago with billions of pounds spent on it. Fabrication facilities exist all over the world and permit the production of ICs at a very low unit cost. For example, both a bipolar operational amplifier μA741CP for use in an analogue circuit and a quad 2-input NAND gate IC from the 74 series CMOS logic facility (74AC00 in standard d.i.l. plastic package) for use in a digital circuit can be purchased for a few tens of pence. In addition, the reliability of solid-state electronic components (e.g. an *npn* transistor in CMOS or TTL family) is greatly superior to that for mechanical components (e.g. relays or circuit-breakers). This recent advance in processor technology has led to a considerable demand for small sensors or *microsensors* that can fully exploit the benefits of IC microtechnology.

If we represent the set of all sensors Ω_1 by a region in space (Figure 1.11), then microsensors are a subset of this which we can represent as an inner region, Ω_3. For the purposes of this book I have defined a microsensor as a device that has a sensing element with a physical dimension on the sub-millimetre (≤ 100 μm) scale. This excludes sensors that measure either small quantities (e.g. a micrometer) or to a high precision (e.g. a laser range finder). The bulk of microsensors may be defined as solid-state sensors (Ω_2) such as the silicon photodiode, phototransistor etc. (§6.3). This class of sensor is fabricated using microelectronic thin-film technology as described in Chapters 3 and 4.

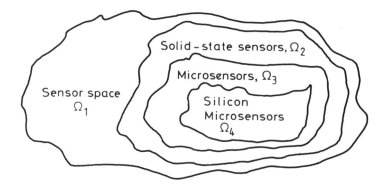

Figure 1.11 Classification of sensor space. After [1.3].

Table 1.6 shows the number of solid-state sensors sold in 1987 together with their value. The list is not complete as it excludes many types of mechanical sensors and biosensors. However, in 1987 1,096 million solid-state sensors were sold worth £600 million - and this only accounts for the Japanese market! The world sensor market was worth £7,000 million in 1989 and is estimated to be worth £25,000 million by the year 2000. The low cost of solid-state sensors coupled with their high reliability and small size has created many applications.

A large proportion of solid-state sensors have thin film sensing elements and so may be classed as microsensors (Ω_3). There is also the latest addition to solid-state sensors, namely silicon microsensors (Ω_4). These sensors are fabricated by chemical etching of either bulk silicon or thin surface layers of polysilicon (§3.3, 3.4). Silicon microsensors often have small overall dimensions and necessitate the integration of some of the processor (§11.1).

Figure 1.12 shows the principal elements of an information-processing system - a sensor, processor and actuator and the application of integration. Integration of the sensor and part of the processor is often desirable as the characteristics of the sensor can be improved by, for example:

- linearising sensor output;
- compensating for temperature or humidity;
- noise reduction.

Table 1.6 Japanese market for some solid-state sensors in 1987.

Sensor type	Class of sensor	Units sold (million)	Value (£M)
Thermal	Thermal	434	87
Optical	Radiation	180	108
Displacement	Mechanical	189	140
Magnetic	Magnetic	218	120
Gas/humidity	Chemical	10	14
Total:	-	1,031 million	£469 million

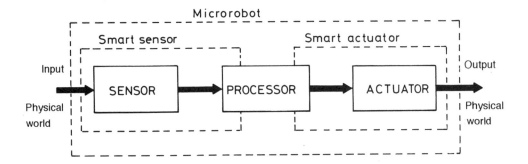

Figure 1.12 Schematic definition of electronic devices produced by the integration of the elements in a measurement system.

A sensor with a partly or totally integrated processor is called an *integrated* or *smart* sensor. Similarly, an actuator with part or total integration of the processor could be called an integrated or smart actuator [1.3]. A detailed discussion of smart sensors may be found in Chapter 11. Full integration of the sensor, processor and actuator is possible although the major problem at present is that the actuator power tends to be rather low. Nevertheless, we may well see integrated measurement systems or micro-instruments in the future, and with more sophisticated processing architecture (e.g. closed-loop control) devices that could be thought of as microrobots or micro-automata.

It may soon be possible to make autonomous microrobots that can be programmed for a given task and propelled via a microactuator to carry them out. Clearly, there is some danger in letting loose microrobots in the environment should a fault occur. For example, a microrobot programmed to remove a blood clot in an artery may start attacking some vital organs in the body. Yet to me the real danger comes when microrobots are given high level intelligence or even the ability to replicate themselves. A mechanical virus may be a reality of the future!

Suggested further reading

Readers are referred to the following texts that provide some background information for this introductory chapter on sensors:

Usher MJ: *Sensors and Transducers* (1985). Published by MacMillan Publishers Ltd, London. ISBN 0-333-38710-4. (163 pages. Sound introduction to sensors)

Hauptmann P: *Sensors: Principles and Applications* (1993). Published by Prentice Hall International (UK) Ltd, UK. ISBN 0-13-805789-3. (215 pages. Good overview with chapter on silicon sensors)

References

1.1   From UKSG presentation by Reed Exhibition International, 225 Washington Street, Newton, MA, USA, June 1992.

1.2   Quality vocabulary, *British Standards Institute*, BS 4778: Part 1: 1979 or part of ISO 8402-1986.

1.3   Middelhoek S and Audet SA (1989) *Silicon Sensors*, Academic Press Inc., San Diego, USA, 376 pp.

Problems

1.1   State the principal forms of signal energy that may be used to classify the different types of sensors. Give two examples of sensors in each class.

1.2   Briefly describe the ideal characteristics of a sensor. Explain why you think these characteristics are ideal and desirable in engineering applications.

1.3   Equation (1.6) describes the characteristic behaviour of an ideal sensor. Derive the output $y(t)$ to a sinusoidal input signal of height h and angular frequency ω. Sketch a graph of the variation in the gain of the sensor with frequency, and indicate on the graph the expected output of a real sensor.

1.4   An actuator is designed to have both position and velocity sensors. Draw a possible control diagram showing feedback control of both position and velocity. Derive the transfer function of your control system.

2. Sensor Signals and Interfacing

Objectives

☐ To introduce signal processing techniques

☐ To consider the demands of processing sensor signals

☐ To present some interfacing standards

2.1 Introduction

The poor sensitivity of the sensing element in self-generating sensors often limits their electrical output to a rather low level. This can be due to the low energy of the physical input signal. In such sensors the principle of operation may involve the injection or removal of a small amount of charge q (e.g. a photodiode in its photovoltaic mode, §6.3.2) or a change in the potential (e.g. a thermocouple, §5.2.2). In contrast, the electrical output from an active or modulating sensor (e.g. a platinum resistance thermometer, §5.4.1) is often higher, less noisy and thus generally easier to process. As discussed in Chapter 1, it is the job of the processor to take the signal from the sensor and, where necessary, to amplify, condition or interface it to some other device (e.g. a microcomputer via a serial transmission line or instrument bus).

Processors that convert electrical properties or signals in the sensing element are sometimes called electrical transducers or electrical sensors but this terminology can be misleading and is not widely used. It is often necessary to measure the resistance, capacitance or inductance of a sensing element as it is modified by the measurand (e.g. the temperature or magnetic field). However, the device is still referred to as a sensor, in the examples here, as a thermal or magnetic sensor and not as a resistance transducer. Consequently, a sensor may be made up of a sensing element and an electric transducer in which case the electric transducer can be regarded as part of the sensor and not the processor.

The processing of sensor signals can take many forms. Some examples are given in the chapters describing the different types of sensor, e.g. thermal or magnetic. Often the processing takes the form of a basic analogue or digital circuit to condition the electrical signal for subsequent use. Clearly the details of the circuit depend upon the nature of the sensor and application desired. Optimum results are obtained where a sensor is not treated as separate from its processor but where the properties of the sensor *system* are considered. After all it may be the interface circuit or the processor which limits the response of a sensor system rather than the physical properties of the sensing element alone. For example, in an open loop system with a sensor s and processor p in series the transfer function of the system is $H_s(s).H_p(s)$ where $H_s(s)$ is the transfer function of the sensor, and $H_p(s)$ is the transfer function of the processor. Thus the band-width of the sensor system could be limited by either the sensor or the processor. It is the job of the engineer to ensure that the design of the processor matches the sensor to the particular application required.

The use of microsensors results in smaller signals to process and this puts additional demands upon the performance of the processor. For example, a micromachined silicon resonant pressure sensor (§7.6.4) may have a capacitance of a few femtofarads.

When considering the different types of sensors, it is quite common to find the same electrical parameters being measured (e.g. resistance, capacitance or inductance) and a need for the signals to be converted (e.g. charge-to-voltage or current-to-voltage conversion). Consequently, this chapter first introduces some of the basic types of electrical sensing elements which produce a signal commonly fed into signal conditioning or other processing devices. Then some common analogue and digital interface standards are discussed together with the need for a sensor standard.

2.2 Measurement of Resistance, Capacitance and Inductance

2.2.1 Resistance

There are many types of sensor in which the measurand causes a change in the electrical resistance R of a sensing element (e.g. magnetoresistors, §8.3 and chemoresistors, §9.2). In the case of a linear resistance sensor the output R_x can be expressed in terms of a baseline resistance R_0 (i.e. when the measurand signal is zero) and its fractional change[1] in value x, where

$$R_x = R_0(1+x) \tag{2.1}$$

For negative values of x, x ideally takes a value of between 0 and -1 so that the sensor output covers the full resistance range with the resistance falling to zero when the input signal is at a maximum. In practice, linear potentiometers that act as position sensors (§7.3) will cover the ideal range of x = [0,-1], but more often the

---

[1] In some sensors (e.g. a PTAT) the model becomes $R_x=xR_0$.

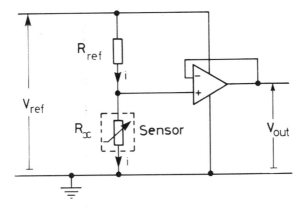

Figure 2.1 Potential divider circuit (when x is large) with buffer amplifier.

value is much lower in the range of 10^{-1} down to 10^{-5} (e.g. a pellistor, §9.6.2). Thus the exact design of the interface circuit depends upon the expected range (and sign) of x.

When large changes in resistance occur ($|x| >> 1$), a simple potential divider or a bridge may be used to process the sensor signal. Figure 2.1 shows a simple potentiometer in which a standard reference resistor R_{ref} is placed in series with the sensor of unknown resistance R_x the output of which is fed into a voltage follower of unity gain to buffer the signal. Application of a constant input voltage V_{ref} leads to a current i flowing through both of the resistors. The voltage output V_{out} can now be related to the sensor resistance R_x by,

$$V_{out} = V_{ref} \frac{R_x}{(R_x + R_{ref})} \tag{2.2}$$

The system gain A is defined here as V_{out}/R_x and is only approximately linear when $R_x << R_{ref}$ and in this case

$$A = \frac{V_{ref}}{(R_x + R_{ref})} = \frac{V_{ref}}{R_0(1+x)+R_{ref}} \tag{2.3}$$

The sensitivity S of the potential divider is given by

$$S = \frac{dV_{out}}{dR_x} = \frac{V_{ref} R_{ref}}{(R_x + R_{ref})^2} \tag{2.4}$$

Figure 2.2 shows a plot of the sensitivity of the potential divider against the sensor resistance R_x. Clearly, the sensitivity rolls off quickly when R_x approaches R_{ref}. The maximum sensitivity occurs at $R_x << R_{ref}$ but the output is a non-linear function when x is large. For a linear sensor, the maximum sensitivity can be calculated from equations (2.1) and (2.4) with $V_{ref}=V_0$,

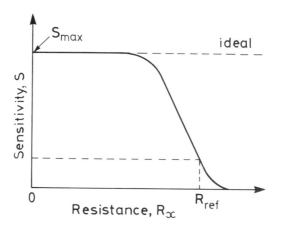

Figure 2.2 Sensitivity of a potential divider.

$$S_{max} = \frac{V_{ref}}{R_0(2+x)^2} \qquad (2.5)$$

In applications where the fractional change in sensor resistance x is small, e.g. up to 1%, the error in the sensitivity calculated by equation (2.5) may be acceptable with a maximum value of only 0.5%. In other applications, where x is large it may be necessary to use a larger reference resistor or to compensate for the non-linear output of the sensor signal.

Potential dividers are very simple to use but they do suffer from several disadvantages. First, they are not very sensitive to small changes in x. Secondly, the output voltage V_{out} depends not only upon the input resistance of the next device but more importantly on the temperature-dependence of the sensor itself. The use of an active divider where the reference element is a duplicate of the sensing element (but does not see the measurand) can obviate this disadvantage (§11.2). The problem of a poor sensitivity may be overcome by using a Wheatstone bridge, as illustrated in Figure 2.3, rather than a potential divider. The Wheatstone bridge consists of a constant voltage supply V_{ref} across a network of four resistors R_1, R_2, R_3 and R_x. The bridge is normally run in the balanced mode (i.e. null mode) but can be run in the unbalanced or deflection mode. The output voltage V_{out} of the bridge may be balanced, manually or automatically, via a variable precision resistor R_1. The resistors R_2 and R_3 set the gain of the network. The balance condition is independent of any internal resistance of the voltage source or the load but they do determine the precision with which balance is maintained.

The output voltage V_{out} of the bridge is given by

$$V_{out} = V_{ref} \cdot \left(\frac{R_x}{R_3 + R_x} - \frac{R_1}{R_1 + R_2} \right) \qquad (2.6)$$

At balance $V_{out} = 0$, so

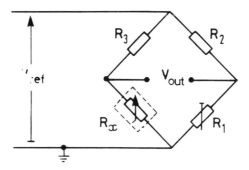

Figure 2.3 Wheatstone bridge arrangement.

$$\frac{R_x / R_3}{1 + R_x / R_3} = \frac{R_1 / R_2}{1 + R_1 / R_2} \tag{2.7}$$

Consequently we must have the familiar general condition that,

$$\frac{R_x}{R_3} = \frac{R_1}{R_2} \tag{2.8}$$

The sensitivity of the bridge circuit may be defined as the bridge output voltage for a change of sensor resistance divided by the supply voltage. For an ideal bridge supply and detector, the sensitivity is

$$S = \frac{x}{(2 + R_1/R_2 + R_2/R_1)} \tag{2.9}$$

with a maximum value of $\frac{1}{4}x$ when $R_1/R_2 = 1$.

The bridge is often used in the unbalanced mode when $|x| \ll 1$. In this case, the resistances R_2 and R_3 are set to the baseline value R_0 and R_1 becomes a dummy sensor of resistance R_0, see Figure 2.4. The output voltage V_{out} is given by

$$V_{out} = \frac{V_{ref}}{2}\left(1 - \frac{1}{(1-x/2)}\right) \tag{2.10}$$

So the output voltage is approximately linearly related to the fractional change in resistance when x is small,

$$V_{out} \approx \frac{V_{ref}}{4}.x \tag{2.11}$$

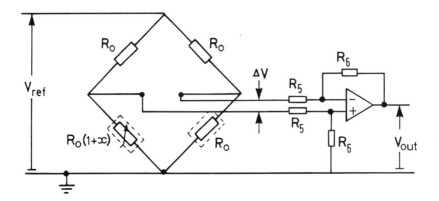

Figure 2.4 Unbalanced Wheatstone bridge circuit with compensation element.

The use of a dummy sensor as R_1 in the bridge gives excellent compensation for sensor shortcomings such as temperature dependence. This is particularly important for some chemical sensors, such as the pellistor (§9.6.2), and for strain gauges (§7.6.5).

The output from such a bridge can be (pre)processed further, for example amplified using a single differential op-amp as shown in Figure 2.4 to produce a respectable voltage. The output voltage of the system now becomes,

$$V_{out} = -\frac{R_6}{R_5} \Delta V \approx -\frac{R_6}{R_5}\left[\frac{V_{ref}x}{4}\right] \text{ when } x \ll 1 \tag{2.12}$$

Although a bridge circuit may null the voltage across the bridge, the performance of the signal processing circuit must always be considered. High input impedances are desirable and a low temperature sensitivity of the circuit. External resistances should be matched to maximise the Common Mode Rejection Ratio (CMRR). It is customary to use an instrumentation amplifier to read the bridge

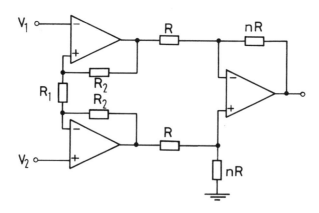

Figure 2.5 Cross-coupled follower instrumentation amplifier.

output due to its low drift, high common-mode rejection, and high input impedance (see Figure 2.5). Further details on bridge circuits may be found in [2.1]. These devices are commercially available at a low cost.

2.2.2 *Capacitance*

In some types of sensor the measurand can cause a change in the reactance (either capacitance or inductance) of a device rather than its resistance. The capacitance C of a device is a measure of the amount of charge q stored for a given voltage V, whence

$$C = \frac{q}{V} \tag{2.13}$$

The capacitance of a device depends upon the geometrical arrangement of its electrodes and the dielectric material employed between them. For a long planar, parallel-plate capacitor, the capacitance C is given by

$$C = \varepsilon_0 \, \varepsilon_r \, \frac{A}{d} \tag{2.14}$$

where ε_0 is the permittivity of free space (8.85 pF/m), ε_r is the relative permittivity of the material between the plates (air is about 1.0, and water is about 100), A is the area of the plates and d their separation. Consequently, any phenomenon that changes the dielectric constant, area or separation of the capacitor plates will cause a change in the capacitance δC. This variation can be defined by the total differential formula,

$$\delta C = \left. \frac{dC}{d\varepsilon} \right|_{A,d} \delta\varepsilon + \left. \frac{dC}{dA} \right|_{\varepsilon,d} \delta A + \left. \frac{dC}{dd} \right|_{\varepsilon,A} \delta d \tag{2.15}$$

For example, the mechanical displacement of one electrode relative to the other could cause either the separation to change by δd or the effective area by δA. This principle is exploited in a variety of mechanical microsensors (e.g. §7.3, 7.6) which use the capacitance change in a simple read-out circuit. Alternatively, the dielectric constant of some materials varies with temperature or chemical species and can be used to form the basis of a sensor (§9.3).

A sensor with a varying capacitance can be measured by placing it in a capacitive voltage divider. The output voltage V_{out} can then be related to the input alternating driving voltage V_{ref}, sensor capacitance C_x and reference capacitance C_{ref} by

$$V_{out} = V_{ref} \, \frac{C_x}{(C_x + C_{ref})} \tag{2.16}$$

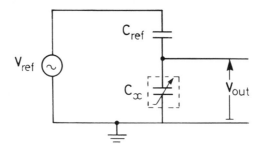

Figure 2.6 Voltage divider for a capacitive sensor.

Clearly, this represents an arrangement like the resistive potential divider (§2.2.1) and again the output is non-linear. Figure 2.6 shows the basic arrangement of a capacitance voltage divider.

The capacitance can be more accurately measured in a bridge circuit. Figure 2.7 shows two bridge arrangements, one which measures an unknown capacitance C_x in series and the other in parallel with an unknown resistance R_x. Ideally, the capacitive sensor resistance R_x is zero (series) or infinite (parallel) but in practice the value must be measured. Capacitive bridge circuits are commonly used and again a dummy sensor can be used in one arm to compensate for temperature drift. The balance conditions are

$$\text{Series:} \quad C_x = \frac{R_2}{R_4} C_1 \text{ , and } R_x = \frac{R_4}{R_2} R_1 \tag{2.17}$$

$$\text{Parallel:} \quad C_x = \frac{R_2}{R_4} C_1 \text{ , and } R_x = \frac{R_2}{R_4} R_1$$

When the capacitance of the sensor is very small, the capacitance of the leads (~ 10 pF/m) may become a problem. In these cases, a guard-ring arrangement may be

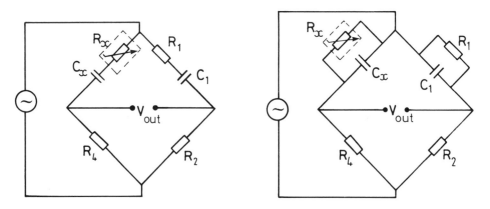

Figure 2.7 AC capacitance bridges.

used to reduce the effect of the cable capacitance. In the case of silicon mechanical microsensors (e.g. §7.6), the capacitances are so small (~ fF) that special attention must be paid to the circuit technology. Possible solutions are to integrate the bridge circuit and sensing element on a single chip and/or to use CMOS technology to produce interconnects, FET amplifiers etc. with low input capacitances and high input resistances.

2.2.3 Inductance

The self-inductance L of a device is

$$L = N\frac{d\Phi}{di} \qquad (2.18)$$

where N is the number of wire turns, Φ the magnetic flux and i the current flowing through the wire. The magnetomotive force (*mmf*) is Ni and so the inductance may be related to the reluctance R_L by

$$L = \frac{N^2}{R_L} \qquad (2.19)$$

where $R_L = \text{mmf}/\Phi$. The magnetic reluctance of a system is analogous to electrical resistance. The total reluctance of a coil, which has a cross-section of A and length l with μ_0, and μ_r the relative permittivity of the air and the iron core, is

$$R_L = \frac{l}{\mu_0\mu_r A} + \frac{l_0}{\mu_0 A_0} \qquad (2.20)$$

where the second term includes the path of the field lines through the air l_0 with cross-section A_0.

Most inductive sensors change the reluctance R_L of a magnetic circuit. There are two main types: variable gap sensors which modify the length of the air gap, l_0 and moving core sensors which modify the magnetic permittivity μ. Inductive sensors are commonly used for the measurement of displacement (§7.3.2) and often run in a differential mode to help reduce the effects of stray magnetic fields and changes in temperature. Inductive sensors can be highly linear (~ 0.5%) but large inductances of ~mH can only be achieved in physically large devices.

The Linear Variable Differential Transformer (LVDT) is based on the mutual inductance between a primary winding and each of two secondary windings as a ferromagnetic core moves inside it. Figure 2.8 shows the circuit diagram for a LVDT. Provided the load resistance on the secondary is high, the primary current is nearly a constant regardless of core position and so the voltage output is

$$V_{\text{out}} = (M_1 - M_2)\frac{di_p}{dt} \qquad (2.21)$$

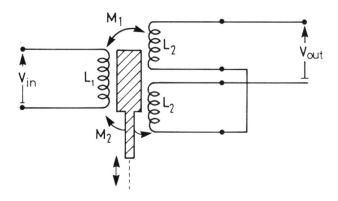

Figure 2.8 General arrangement of a LVDT.

where the difference in mutual inductance $(M_1 - M_2)$ shows a fast and linear change on both sides of the central magnetic core position. In practice the output voltage depends upon the excitation frequency and on the load resistance (typically 10 kΩ). In addition, the output voltage is often not zero at the centre position due to a difference in the windings and its accuracy can be temperature-dependent via, for example, the primary coil resistance fluctuating. Nevertheless, a typical LVDT run at 5 V, 2 kHz can measure displacements down to only a few microns over a range of 1 mm or so.

2.3 Interface Electronic Circuitry

The electrical output from a sensor usually requires some form of processing or conditioning before it can be used by another device. The signal processing circuitry which lies between the sensor and a data system or actuator is usually called the interface electronics. Figure 2.9 shows the basic arrangement of a sensor system. The interface electronic circuit may simply convert a variation in resistance, capacitance or inductance to a voltage signal as described previously (§2.2) but it also has other possible functions, see Table 2.1.

Table 2.1 Common functions of sensor interface electronics.

Function	Examples
To amplify	Non-inverting voltage op-amp circuit using bipolar IC technology.
To reduce noise	Passive *RC-CR* network as band-width filter, digital filter.
To supply power	Bias voltage in a thermodiode.
To compensate	PTAT thermal sensor, active bridge configuration.
To control	Self-test in a smart sensor.
To transmit	4 to 20 mA output standard, RS-232 standard.

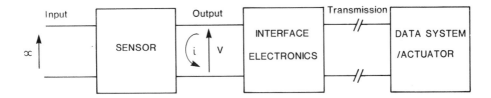

Figure 2.9 Block diagram of a sensor system showing interface electronics.

The major role of the interface circuit is usually to amplify the raw signal to a level where it is of practical use. Conventional analogue circuitry is often used such as bipolar op-amps to form inverting or non-inverting amplifiers. The integration of this circuit with a silicon sensor may be desirable when some signal buffering is also needed. For instance, it may be necessary to reduce the output impedance of the sensor via a voltage follower.

A secondary role of the interface electronics is to reduce noise by suitable filtering. This may require the addition of a low-pass, high-pass or band-pass filter which will also limit the dynamic range of the sensor output. The filter may be included in the main amplifier circuit, e.g. an integrating amplifier, where desirable.

Third, the interface electronics will supply the power and/or carrier signal to modulating sensors. Thus the sensor performance is often determined by the stability of the interface electronics in producing a stable power supply. This is less important in the case of a bridge circuit where, to first order, fluctuations of the power supply are irrelevant.

Fourth, a role of interface electronics may be to correct for some deficiency in the sensor performance (§10.2). For example, a circuit may be introduced to linearise the output of the sensor against the input. A linear output is highly desirable (§1.3) and so some effort may go into the analogue electronics to ensure that this occurs. Examples are the PTAT thermal sensor (§5.6) which produces an output signal that is proportional to the absolute temperature. More sophisticated interface electronics may include log-amplifiers to linearise, or perhaps an analogue structural design to compensate for temperature drift (§5.6). These latter sensor systems may be regarded as having a certain intelligence, and, if integrated with the sensing element, they would be called smart sensors (§11.1).

Fifth, the interface electronics may have some facility to produce a control signal, e.g. a low supply voltage level warning signal or self-test mode. Although these are possible in an analogue electronic circuit they are more common, and easier to implement, using a digital data processing system. Finally, the interface electronics may prepare the signal for transmission to another device. The principal forms of data transmission are, analogue current or voltage, frequency modulation, and digital transmission.

It is quite acceptable to use a voltage signal (e.g. 1 to 10 V DC) for transmitting signals over a short distance, e.g. along a ribbon cable to an analogue-to-digital converter or multiplexer. However, in situations where an <u>analogue</u> signal needs to travel a longer distance then a current loop is preferable. The 4 to 20 mA current loop is an industrial standard and the sensor signal should lie within this range. When no current is measured it is indicative of a failure of either the cable, the interface electronic circuit or the sensor itself.

2.4 Frequency Modulation

In some types of sensor it is useful to convert the sensor signal to a frequency before transmission or read-out. The conversion of an electrical property or signal into a frequency signal can be advantageous. For example, the conversion of an analogue voltage to a frequency, Figure 2.10(a), permits precise measurement since commercial frequency counters can measure to one part in tens of millions.

Some sensors require a frequency signal for modulation. For example, resonant silicon mechanical sensors (e.g. §7.6) operate at baseline frequencies of kHz to MHz and can provide a high level of sensitivity to measurand, e.g. pressure. Similarly, piezoelectric and SAW devices are resonant acoustic devices which, when coated with chemically-sensitive materials, can be sensitive at the ppb level (§9.7). In situations where an array or a set of sensors are employed the analogue input can be multiplexed to reduce the amount of cabling required. This is useful when reading out from a collection of discrete sensors, perhaps a pressure sensor, temperature sensor and humidity sensor. The multiplexing reduces the complexity of the sensor system by reducing the number of interconnecting cables and thus cost. This system may be used to process larger arrays of sensors (§12.1), although the demand for high serial processing rates may require the use of other devices. For example, shift registers are used in CCD video cameras to obtain MHz rate data streams (§12.2.2).

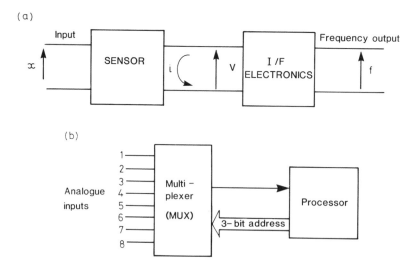

Figure 2.10 (a) Voltage-to-frequency converting interface; (b) multiplexing interface.

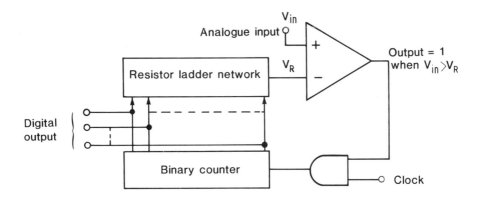

Figure 2.11 Single ramp ADC.

The Analogue-to-Digital Converter (ADC) is widely used to provide a digital signal which can then be transmitted along cables with higher data rates and with more noise immunity. Figure 2.11 shows the simplest form of an ADC which is a single ramp ADC. The circuit works by comparing the analogue input voltage V_{in} with an accurately timed ramp signal generated by the ladder network. Many low-cost ADCs still operate on this principle, such as the Ferranti ZN245, but successive approximation ADCs are now more common. Table 2.2 lists 4 monolithic and 3 hybrid ADCs representative of commercial ADCs [2.3].

Table 2.2 Some commercially available ADCs.

Type	Makers	Resolution (bit)	Conversion time (µs)	Supply (V)
AD570	Analog Devices	8	25	+5, -15
TSC7109	Intersil	12	33 ms	+5
ADC0808	National	8	100	+5
MC10317	Motorola	7	30 ns	+5, -5
AD579	Analog Devices	10	2.2	+5, ±15
AD574	Analog Devices	12	25	+5, ±15
ADC71	Burr-Brown	16	50	+5, ±15

2.5 Digital Transmission

2.5.1 Serial interfacing

The preparation and transmission of digital signals by interface electronic circuitry is common and the cost is usually offset by the advantages given above of higher transmission rates and lower noise levels. For example, Table 2.3 shows the typical transmission characteristics of cables supporting signals from digital interface electronic circuitry. It can be seen that the cost of optical fibre cabling is potentially advantageous when transmitting large volumes of data at high speed. Digital information is usually exchanged between a device (e.g. sensor) and a microcomputer by either a serial or a parallel communications port. In theory a serial digital interface requires only one or two wires and is significantly cheaper than a parallel (e.g. 8 wire) interface. Moreover, a dual-wire transmission can use commercial communication facilities such as a standard telephone cable.

Serial interface standards have been defined by the Electronics Industries Association (EIA). The most commonly used interface for microcomputer systems is the RS-232 standard, while the RS-422 and RS-423 standards are used where longer transmission lines or higher data flow-rates are required. Figure 2.12 shows the basic configuration of these three serial interfaces. The bit streams are generated by standard drivers (e.g. a Motorola MC1488 or TI SN75188) and collected by standard receivers (e.g. a Motorola MC1489 or TI SN75189). The RS-232 connector is a 25 pin D connector in which most of the lines are control lines. This rather complicated arrangement has arisen from its original use in exchanging information with modems, printers and plotters rather than sensors. In some ways this spoils the ideal of a simple, low-cost transmission line for a remote sensor. In practice the recommended line length (load dependent) for the RS-232C interface is about 100 m and the maximum data rate is 20 k*baud* (20 kbits/s in a binary channel). Inevitably the serial interface is designed to connect one device (e.g. a single sensor or set of sensors via multiplexing) to one microcomputer.

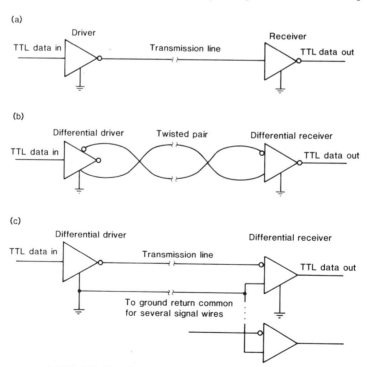

Figure 2.12 (a) RS-232 (Single-ended), (b) RS-422 (Balanced differential) and (c) RS-423 (Unbalanced differential) serial interfaces.

Table 2.3 Transmission of signals by cables [2.2].

Cable	Max. rate (bits/s)	Weight/length (g/m)	Cost/length ($/m)	Max. temp. (°C)
Twisted pairs	7 MHz	8.6 to 3.2	1.00 to 0.64	80
Ribbon cable	5 MHz	0.10	0.12	105
Coaxial cable, RG-58/U	20 MHz	48.0	0.44	75
Computer cable, RG-62/U	40 MHz	54.8	0.61	80
Fibre optics				
single mode	10 GHz	0.13	0.25	80 to 200
multimode	200 MHz	0.21	0.17	80 to 200

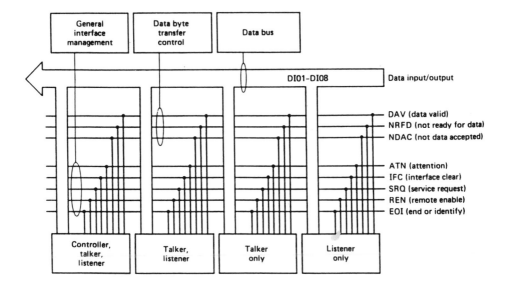

Figure 2.13 General arrangement of the GPIB.

2.5.2 GPIB (IEEE 488) interfacing

Manufacturers have developed an interface system specifically for digitally controlled instruments. The communication and control strategy is called the General Purpose Instrument Bus (GPIB) and this was standardised by the IEEE in 1978 [2.3] as the IEEE-488. The bus arrangement permits a single processor (e.g. a PC) to control as many as 20 other devices using only one interface. The GPIB uses an asynchronous hand-shake to pass data and so transmission rates much higher than the RS-232C are possible (i.e. up to 2 Mbits/s). The GPIB is a set of 25 lines, 8 to carry data, 8 to act as control lines and earths. Any device connected to the bus may send (talk) or receive (listen) data but one device must act as a controller, see Figure 12.13. Interface cards are commercially available for PCs which could contain an Intel 8291 talker/listener, 8292 GPIB controller and two 8293 GPIB bus transceivers. This permits the PC to manage the bus system via

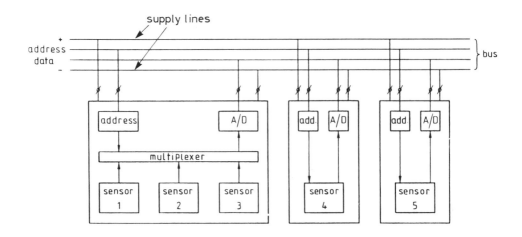

Figure 2.14 A possible sensor bus interface. From [2.4].

software. Although this interface system provides enormous flexibility of design, e.g. sensors can be easily added or removed from the bus, the interface cards are rather expensive and so such a system is not suitable for low-cost sensing applications. It is only suitable for short range communications (up to 20 m).

2.5.3 *Sensor interfacing*

The potential advantages in the integration of the signal processor with the sensor, such as in lower cost, lower weight, smaller size etc., are discussed elsewhere (§1.5, §11.1). Yet there is still a considerable need to develop a general purpose smart interface device that sits between a data bus and the individual sensors. Figure 2.14 shows a possible bus interface structure that only uses 4 serial digital lines instead of the 25 in the IEEE-488 standard [2.4]. Sensor subsystems can then be readily added or removed from the bus structure. These subsystems could be individual sensors, multiplexed discrete arrays, or array sensors. To date there is no bus standard specifically for sensor systems. However, the next decade should see the arrival of a new family of smart interfaces using a simple bus standard.

Suggested further reading

Readers are referred to the following texts that provide some background information for this chapter on sensor signals and interfacing:

Bannister BR and Whitehead DG: *Transducers and Interfacing* (1986). Published by Van Nostrand Reinhold, UK. ISBN 0-442-31742-5. (112 pages. Basic introduction to interfacing)

Sheingold DH: *Transducer Interfacing Handbook* (1980). Published by Analog Devices Inc., USA. ISBN 0-916550-05-2. (A detailed guide to analogue signal conditioning)

References

2.1   Pallas-Areny R and Webster JG (1991) *Sensors and Signal Conditioning*, John Wiley & Sons, New York, 398 pp.

2.2   Course notes from *Microsensors and Microactuators*, Oxford/Berkeley Summer Engineering programme, University of Oxford, UK, July 1990.

2.3   IEEE-488 (1978) *Standard digital interface for programmable instrumentation*, New York, USA.

2.4   Middelhoek S and Audet SA (1989) *Silicon Sensors*, Academic Press, San Diego, USA, Ch. 8, p358.

Problems

2.1   Design a basic op-amp circuit for three different types of electric transducers or electrical sensors. Discuss the advantages and disadvantages of this circuit when compared to a bridge circuit with a dummy sensor arm.

2.2   Describe what effect the power supply and deflection meter have on the performance of a Wheatstone bridge circuit. Derive equations, where possible, to illustrate your answer.

2.3   State the functions that an interface electronic circuit may perform in a sensor system. What do you think are the most important effects to compensate for when using (a) a thermocouple (b) a photodiode?

2.4   What are the RS-232C and IEEE-488 standards? Give an example of a sensor application where you think that a serial interface is preferable to a bus interface and vice versa. Briefly explain why in each example.

3. Conventional Silicon Processing

Objectives

- [ ] To discuss materials processing for microsensor fabrication
- [ ] To consider silicon planar technology
- [ ] To describe bulk and surface micromachining
- [ ] To review silicon bonding technology

3.1 Introduction to Silicon Processing

A major advance in engineering is often realised when an initial advance in the science base is followed by an enabling advance in the technology base. This observation particularly applies to the field of microsensors, where there is a considerable demand for microsensor technology to be compatible with current microelectronic technology.

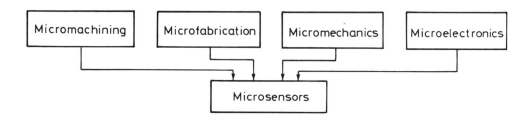

Figure 3.1 Basic engineering area involved in microsensor technology.

The successful engineering of a microsensor needs inputs which rely on the development of appropriate fabrication processes and understanding in four major areas: *micromachining, microfabrication, micromechanics,* and *microelectronics,* see Figure 3.1. The basics of the first two processes are described specifically within this and the next chapter, while the principles of the latter two disciplines are described in the chapters on sensor types (Chapters 5 to 9), interfacing (2), smart sensors (11) and microsensor array devices (12).

Microsensor technology is challenging because it involves the processing of a very wide range of materials. This can place a severe demand upon the sensor technology. In the manufacturing of microsensors, conventional silicon planar microelectronic technology has been developed to accommodate the processing of both passive and active materials. A passive material is one that does not play an essential role in the sensing mechanism (e.g. an SiO_2 insulating layer in a pressure sensor, §7.6.4) in contrast to an active material which does (e.g. a metal oxide layer in a chemical sensor, §9.2.2).

Both this chapter and the next discuss the basic materials' processing required to make microsensors. This is commonly known as *micromachining* and *microfabrication.* This chapter describes the principles of the wet etching of microsensors in bulk silicon or thin surface layers, together with some basic bonding technology. The next chapter discusses not only some of the more specialised techniques used in micromachining and microfabrication, such as ion beam milling, but also the deposition of some active materials such as ultra-thin Langmuir-Blodgett films, electropolymerised thin films and screen-printed thick films. To begin, a review is made of the basic processing steps in silicon planar IC technology, since these often form the framework of microsensor technology.

3.2 Silicon Planar IC Technology

The fabrication of microsensors usually requires the use of conventional silicon planar IC technology that has been subsequently modified to include some additional processing steps, such as polysilicon deposition.

Monolithic silicon technology was first described over thirty years ago [3.1]. Since then a range of monolithic silicon integrated circuit technologies have been implemented, which can be related to two basic types known as bipolar and MOS. Figure 3.2 shows the families of ICs now available together with their circuit type and isolation geometry. CMOS is one of the most common IC technologies currently used in microsensors. Each of these ten or so types of IC technology has its own specific requirements for materials processing but there are a considerable number of common steps that are of importance here.

Silicon planar IC fabrication generally involves all of the following processes:

- crystal growth and epitaxy;
- oxidation and film deposition;
- diffusion or implantation of dopants;
- lithography and etching;
- metallisation and wire bonding;
- testing and encapsulation.

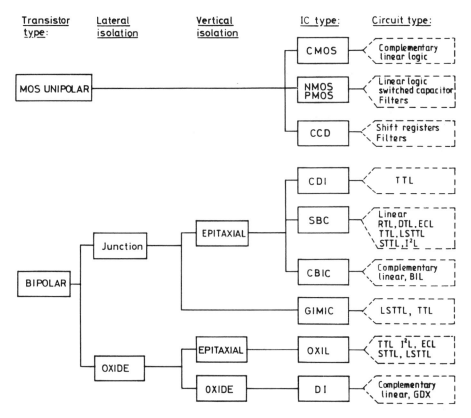

Figure 3.2 Monolithic silicon integrated circuit technologies. From [3.2] (© 1976 IEEE).

In order to give the reader an appreciation of the processes involved, the basic steps are now described to fabricate an *n*-channel Metal Oxide Semiconductor Field Effect Transistor (MOSFET) from a silicon wafer.[1] Figure 3.3 summarises the ten principal processing steps required. The majority of ICs are made from large single crystals of silicon obtained either through Czechralski crystal pulling or, if very pure silicon is required, through a float zone process (see [3.3] for details). Although silicon has lower electron and hole mobilities (1,500 and 450 cm^2/Vs) than gallium arsenide (8,500 and 1,500 cm^2/Vs), it is generally an easier material to process and is more reliable. Consequently, most ICs and sensors are fabricated whenever possible using silicon rather than gallium arsenide wafers .

3.2.1 Diffusion and ion implantation

The MOS transistor is generally made on a silicon wafer that has been doped with an *n*-type material to a depth of about 100 μm followed by a light *p*-type doping to about 10 μm (Step 1 in Figure 3.3). Controlled amount of dopants are introduced into the wafer by either thermal diffusion or ion implantation. In the thermal

---

[1] A detailed discussion of these topics can be readily found in many books on semiconductor devices and VLSI technology, e.g. [3.3] and [3.4].

diffusion of n-type materials, wafers are placed in a furnace and an inert gas containing the desired dopant (e.g. AsH_3 or PH_3) is passed over them. Alternatively, liquid or solid sources can be used to introduce the dopant material. Similarly, a p-type layer can be produced by passing an inert gas carrying, for example, B_2H_6.

In intrinsic semiconductors the diffusion of the dopant can be described by Fick's diffusion equation,

$$\frac{\partial}{\partial t}C(x,t) = D\frac{\partial^2}{\partial x^2}C(x,t) \tag{3.1}$$

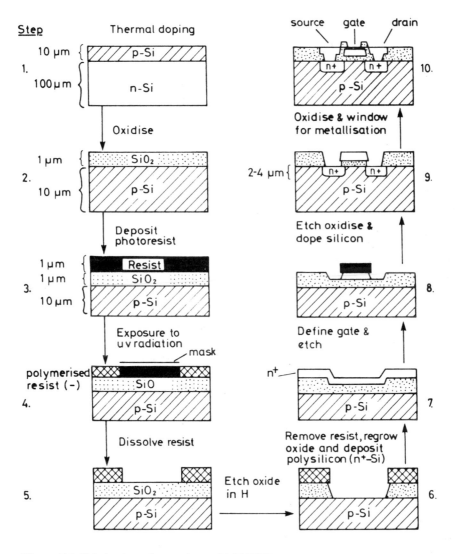

Figure 3.3 Fabrication of an n-channel MOSFET using silicon planar technology.

where $C(x,t)$ is the concentration of the dopant at a distance x into the wafer at time t, and D is a diffusion coefficient whose temperature dependence is given by an Arrhenius equation, namely,

$$D = D_0 \exp(-E_a / kT) \tag{3.2}$$

D_0 is a constant for a given reaction and concentration (see below), E_a is an activation energy with a value that depends upon the transport mechanism (typically 0.5 to 1.5 eV in silicon), k is the Boltzmann constant and T the temperature. The diffusion coefficients of phosphorus, arsenic and boron in silicon at 1,100°C are typically 2×10^{-13}, 1×10^{-13} and 1×10^{-13} cm^2/s respectively [3.5].

In gaseous doping of a silicon wafer, the concentration of the dopant at the surface may be regarded as a constant C_s during the doping process. Assuming that the dopant does not reach the other side of the wafer (i.e. the substrate may be regarded as semi-infinite), then the concentration profile can be readily found from equation (3.1) to be

$$C(x,t) = C_s \, \mathrm{erfc}\, (x / 2\sqrt{Dt}) \tag{3.3}$$

where erfc is the complementary error function and $\sqrt{Dt}$ is called the diffusion length. It should be noted that a Gaussian distribution is produced rather than an erfc function when the doping source is a thin surface film. This is because one boundary condition has changed with the surface dopant concentration C_s falling as the material supply is depleted.

The diffusion coefficient D is independent of the carrier concentration below the intrinsic carrier concentration ($\sim 5 \times 10^{18}$ cm^{-3} for silicon at a temperature of 1,000°C). However above the intrinsic carrier concentration, the diffusion coefficient increases with carrier concentration and generally follows a power law. This is essentially due to the diffusion mechanism becoming vacancy-assisted. Figure 3.4 illustrates the concentration-dependent diffusivity of As and P in crystalline silicon above 10^{18} cm^{-3} [3.3]. The diffusivity of arsenic is roughly proportional to the concentration, while that of phosphorus is roughly quadratic. The exact profile can be found mathematically for a concentration-dependent diffusion coefficient [3.6] but it is common practice to make use of calibration charts to determine the junction depth x_j based upon

$$x_j = k\sqrt{D_s t} \tag{3.4}$$

where k is a coefficient that depends upon the diffusivity region and can take a value from 1.6 (linear dependence) to 0.87 (cubic dependence), D_s is the surface diffusivity and is a function of the dopant material and temperature, and t is the diffusion time [3.7].

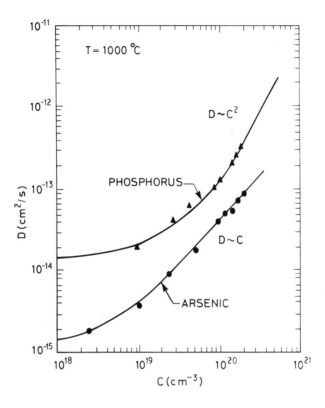

Figure 3.4 Concentration-dependent diffusivity of As and P in silicon. After [3.3, 3.8].

It should always be borne in mind that the concentration profile of the dopant in the semiconductor cannot change abruptly (this would imply a zero or infinite flux) but only gradually so diffusion must take place around the edges of masks, i.e. in the y- and z-directions.

An alternative method to thermal doping is to use ion implantation in which charged ions are accelerated to energies in the range of 10 to 1,000 keV and fired at the semiconductor surface. Figure 3.5 shows an impurity profile of P ions in a silicon target with the beam angle 7° off the normal direction [3.3, 3.8]. The dopant profile will be Gaussian unless, as shown here, there is a beam misalignment which produces an exponential tail. This exponential tail is due to an ion-channelling effect caused by the crystal structure and would not be observed in an amorphous material. The technique is now commonly used with penetration depths of As, P and B typically 0.5, 1 and 2 μm at 1,000 keV in silicon.

3.2.2 Oxidation

An oxide layer of about 1 μm thickness is then grown on the p-Si layer (Step 2 in Figure 3.3). The oxide layer is grown by placing the wafer in a furnace at 1,100°C containing oxygen. The oxygen dissociates, simultaneously reacting with the silicon and diffuses through the growing SiO_2 layer,

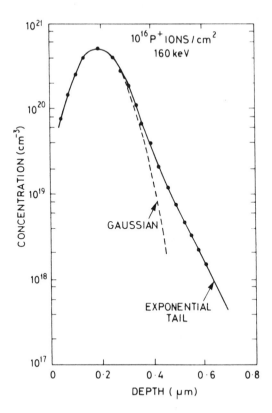

Figure 3.5 Concentration profile of ion-implanted P in crystalline silicon; the ion beam is 7° from the <111> axis to show an exponential tail [3.3].

$$Si + 2[O] \rightarrow SiO_2 \tag{3.5}$$

The mathematical equations that describe this diffusion-reaction process are complex with the problem generally being referred to as the *moving boundary problem;* see Crank for a discussion of this phenomenon [3.6].

Again the complexity of the process dictates the use of a calibration chart to predict the thickness of the oxide layer from the oxidation temperature and time. Oxide layers of up to a few microns can be readily formed in this way but thicker films take a prohibitively long time to form.

The oxide layer is used as a barrier to subsequent doping of areas with *n*-type and *p*-type materials and thus forms an electrically insulating region of the semiconductor wafer.

3.2.3 *Lithography and etching*

Lithography is the process of imprinting a geometric pattern from a mask onto a thin layer (~μm) of material called a *resist* (a radiation-sensitive material). First, a resist is usually either spin-coated or sprayed onto the silicon wafer and then a mask placed above it (Step 3 in Figure 3.3). Second, in optical lithography, UV radiation is used to change the solubility of the photoresist in a known solvent (Step 4 in Figure 3.3). Positive photoresists become more soluble on the exposure to the UV light whereas negative photoresists become less soluble due to a polymerisation process.

After the uncured photoresist has been dissolved away by washing in an organic solvent, the exposed SiO_2 layer is then etched away by an HF solution (Steps 5 and 6 in Figure 3.3) and the remaining polymerised resist is burnt off. Next the gate oxide is formed by another thermal oxidation but to a thickness of only 0.1 µm (Step 7 in Figure 3.3). Note that the processing described here uses a negative photoresist procedure but positive photoresists tend to give better results as the latter do not absorb solvent, swell and lose resolution.

In optical lithography ($\lambda \sim 0.4$ µm) the minimum line-width or resolution is determined by the shadow printing technique used: projection priniting typically has a resolution of 1 to 5 µm, however this has now been reduced to 0.3 to 0.5 µm through the use of harder UV light and better optics. Superior sub-micron resolutions may be achieved through the use of electron, X-ray or ion beams; these are discussed in the next chapter.

3.2.4 Film deposition

The gate electrode is now formed by depositing polysilicon over the entire wafer. Normally silicone is pyrolysed to produce the polysilicon layer:

$$SiH_4 \xrightarrow{\;600°C\;} Si + 2H_2 \qquad\qquad (3.6)$$

The polysilicon layer can be doped *in-situ* through the use of dopant gases or subsequently by diffusion or ion-implantation (Steps 8 and 9 in Figure 3.3). Polysilicon is a good gate material as it has a higher reliability than aluminium. The source and drain channels can then be etched out and doped to a depth of 2 to 4 µm by the thermal diffusion of P_2O_5 or PH_3 gas at a temperature of *ca.* 1,000°C.

3.2.5 Metallisation and wire bonding

Finally, another oxide layer is formed and the windows for metallisation exposed lithographically (Step 10 in Figure 3.3). The metal is then deposited by either physical vapour deposition (i.e. evaporation), chemical vapour deposition or sputtering of a ~1 µm layer. The metal (e.g. Al or Au) forms an Ohmic contact to the source, drain and gate electrode and connections to pads etc. Aluminium generally adheres well to SiO_2 although junction spiking and electromigration are a problem for shallow junctions. In this case, silicides (e.g. $TiSi_2$ or $TaSi_2$) can be used.

The wafer is then diced up using a saw or diamond scribe and the IC mounted onto a package. The electrical connections between the pads and package terminals are usually made by the ultrasonic welding of thin aluminium or gold wire or ribbons.   The wire-bonding process is often crucial to the practical reliability of electronic components as discussed in §10.3.

3.2.6 Passivation and encapsulation

Protection of the IC from the atmosphere is highly desirable for a reliable and robust device. The sensitive areas are often covered up by photoresist or often by a silicon nitride layer. Silicon nitride can be deposited by LPCVD (750°C) or plasma-assisted CVD (300°C) and acts as a good barrier to water. The Si_3N_4 layer can also be used during the fabrication stage to prevent the oxidation of the wafer. LPCVD is often used and can produce Si_3N_4 layer via the following reaction:

$$3SiCl_2H_2 + 4NH_3 \xrightarrow{750°C} Si_3N_4 + 6HCl + 6H_2 \qquad (3.7)$$

Film thicknesses are usually limited to less than 0.2 µm because of high thermally-induced stresses, although it is possible to build up thicker layers through a lamination process. Finally, the IC is encapsulated in some manner, e.g. sealed in a plastic resin or hermetically sealed in a metal case.   This process is highly desirable as it protects the silicon device from the environment and thus improves its reliability (§10.3). Unfortunately, in the case of some microsensors it is not always possible to isolate the silicon structure from the atmosphere, e.g. in a humidity or gas sensor.   A detailed discussion of the design and processing of microsensor packages may be found elsewhere [3.9, 3.10].

3.3 Bulk Silicon Micromachining Techniques

3.3.1 Use of IC processes

The principal processing techniques in the manufacture of silicon microsensors are:

- bulk micromachining;
- surface micromachining;
- silicon bonding;
- active material processing.

Silicon micromachining is the application of silicon planar IC processing techniques to the selective etching of silicon and other films in order to fabricate microstructures (e.g. microsensors and microactuators).   As we shall see, the micromachining of microsensors not only requires all the conventional IC processing techniques described above but additional ones for new passive materials (e.g. amorphous silicon or polyimide) or active materials (e.g. gas-sensitive oxides or polymers). These extra demands are often further increased by our desire to integrate the sensor technology with, say, CMOS circuitry.

We now consider in detail the comparatively new techniques of bulk micromachining, surface micromachining and silicon bonding.

3.3.2 *Isotropic wet etching*

The chemical etching of silicon has been successfully used to make microelectronic devices reliably and cheaply for many years. This technique can now be employed to make structures rather than devices but first consideration must be made of the additional demands put upon the technology. Bulk silicon may be removed by either the chemical etching in a solution (i.e. wet etching) or by producing a plasma (i.e. dry etching). In general the wet etching of silicon produces a better defined structure than dry etching and is more commonly used. Various factors influence the wet etching of silicon such as the etchant, stirring method, temperature, crystal orientation, doping level and crystal defects. When the mass transport of the reactant from the solution to the surface or reactant product back into the solution is slow, the process is said to be diffusion-rate limited and the etch-rate is strongly influenced by stirring. Alternatively, when the surface reaction is slow then the process is reaction-rate limited and the etch-rate is strongly temperature dependent. Consequently, etching apparatus needs to have both a good temperature controller and a reliable stirring facility.

Figure 3.6 shows a typical reactor that can be used to etch silicon wafers [3.11]. The wafer is positioned in the etching solution by a Teflon sample holder. The solution can be stirred using a magnetic spinner and the temperature maintained

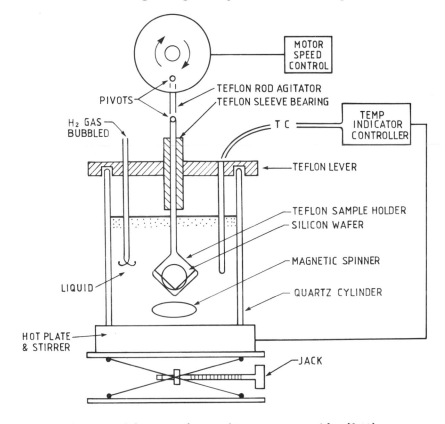

Figure 3.6 Schematic of wet etching apparatus. After [3.11].

by a hot plate.

The first etch solutions to be developed provided an isotropic etch, i.e. the etch-rate was independent of crystal orientation [3.12-3.15]. Isotropic etchants generally consist of a mixture of nitric, hydrofluoric and acetic acids. The nitric acid acts an oxidant, although H_2O_2 on Br_2 can also be used. The hydrofluoric acid then acts as a depassivant and removes the oxide film. The acetic acid acts as a complexant. The exact composition of the $HNO_3:HF:CH_3COOH$ mixture determines the nature of the etching process and the resultant surface. For example, a high HNO_3 concentration produces a diffusion-rate limited etch and gives a smooth surface finish.

Table 3.1 gives a list of commonly used etchant compositions together with their characteristic effect. It should be noted that these etchants generally require specific etch masks such as nitride layers or noble metals, although polymerised negative photoresists and SiO_2 offer limited resistance [3.16].

Table 3.1 Some isotropic etchant compositions. After [3.16].

HNO_3	HF	CH_3COOH	Comments
100	1	99	Removes shallow As implants.
91	9	-	Polishes at 5 μm/min with stirring.
75	25	-	Polishes at 20 μm/min with stirring.
66	34	-	Polishes at 50 μm/min with stirring.
15	2	5	Planar etch.
6	1	1	Slow polishing etch.
5	3	3	Non-selective polishing etch (CP4A).
3	1	8	Selectively etches n^+ or p^+ with $NaNO_2$ or H_2O_2 added.

Figure 3.7 shows a more detailed diagram of the % weights of HNO_3, HF and the diluent (either CH_3COOH or H_2O). The contours represent the etch-rates at different temperatures and from this diagram the desired etchant composition can be selected [3.12]. In practice, there are problems associated with isotropic etching such as the control of the etch-rate and thus the precision of the machining process [3.17]. Because the etch-rate is often diffusion-rate limited, the systematic etching of holes, dimples etc. is quite difficult. Consequently, anisotropy etching is more commonly used than isotropic etching of silicon.

Buffered HF solutions are used to etch isotropically borosilicate and phosphosilicate glasses [3.18] which are used as sacrificial layers in surface micromachining (see next section). Nitride layers can also be etched in phosphoric acid where silicon dioxide can be used as a masking material [3.19].

3.3.3 Anisotropic wet etching

In anisotropic etching of silicon, the removal of silicon occurs at a rate that depends strongly upon the orientation of the crystal lattice structure and the doping level. Anisotropic etchants selectively etch the <100> and <110> crystal

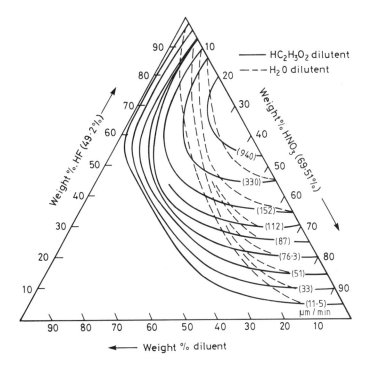

Figure 3.7 Isotropic etching diagram for silicon [3.12]. (Reprinted by permission of the publisher, The Electrochemical Society, Inc.)

orientations while leaving the <111> crystal orientation relatively free. The most commonly used etchants are inorganic alkaline solutions (e.g. KOH, LiOH, NaOH, CsOH) and organic alkaline solutions (e.g. mixtures of ethylene diamine, pyrocatechol and water (EDP) or hydrazine (N_2H_4/H_2O) or NH_4OH).

Table 3.2 lists some of the common alkali etchants together with their typical etch-rates as a function of crystal orientation in silicon and SiO_2. The information is taken from various sources [3.20-3.24]. It may be seen from Table 3.2 that a range of etch-rates can be obtained for the <100> and <110> crystal planes depending upon the degree of anisotropy required. The ultimate choice of an etchant depends upon the overall process requirements but it is beneficial for the SiO_2 etch-rates to be slow. This means that SiO_2 can be used like a nitride layer to act as a good masking material.

The precise mechanism for anisotropic etching is not well understood but is generally believed to be as follows [3.20, 3.21]:

1. Hydroxyl ions bind to either two silicon dangling bonds for the <100> surface or one dangling bond for the <111> surface producing one or two conduction electrons.
2. Two (<100>) or three better-shielded (<111>) back bonds are broken to produce $Si(OH)_2^+$ or $SiOH^{3+}$ together with 2 or 3 conduction electrons.
3. The silicon hydroxide reacts with hydroxyl ions to form orthosilic acid which diffuses from the surface into the bulk solution.

Table 3.2 Anisotropic etching characteristics of crystalline silicon and thermal oxide.

Alkali etchant	Temperature (°C)	Etch rate (μm/h) of			
		Si <100>	Si <110>	Si <111>	SiO_2
Inorganic:					
KOH: H_2O	80	66	132	0.33	-
KOH (42% wt)	75	42	66	0.5	0.34
KOH (57% wt)	75	25	39	0.5	0.62
KOH in H_2O/IPA	80	66	33	3.3	-
Organic:					
EDP[2]	110	51	57	1.25	0.004
N_2H_4: H_2O	118	176	99	11	0.01
N_2H_4: H_2O	110	180	12	-	-
NH_4OH (3.7% wt)	75	24	8	1	0.003
NH_4OH	90	30	-	0.6	-

[2] EDP type S for slow, lower temperature etches (1 litre of ED, 160 g of pyrocatechol, 6 g of pyrazine and 133 ml of water), see [3.25].

4.   Finally, in highly alkaline solution the orthosilic acid dissociates and hydrogen is produced.

Consequently, the overall reaction for <100> etching is given by:

$$Si + 2OH^- + 2H_2O \rightarrow SiO_2(OH)_2^{2-} + 2H_2 \qquad (3.8)$$

Thus the difference in etch-rates may be explained in terms of the different back bond breaking energies and stereo-chemical effects for the crystal planes.

Organic etching is often preferable because it does not produce a K^+ contamination of the silicon wafer. The amine oxidises the silicon with pyrocatechol acting as the complexing agent and pyrazine as a catalyst. The reaction is usually carried out at a temperature of 110°C and is in two stages as follows:

$$Si + 2NH_2(CH_2)_2NH_2 + 6H_2O \rightarrow Si(OH)_6^{2-} + 2NH_2(CH_2)NH_3^+ + 2H_2$$
$$Si(OH)_6^{2-} + 3C_6H_4(OH)_2 \rightarrow Si(C_6H_4O_2)_3 + 6H_2O \qquad (3.9)$$

3.3.4 Anisotropic etch-stops

In order to control the shape of the etched structure, it is often necessary to ensure that the etching stops at a particular position. The etch-rates of alkaline anisotropic etchants are strongly influenced by boron doping of the silicon. Figure 3.8 shows the effect of boron doping on etching of <100> silicon by KOH at 60°C and EDP (type S) at 66°C or 110°C [3.18]. At a doping level of 1×10^{20} cm^{-3}, the etch-rate has fallen by a factor of ~100 for both KOH (10% concentration level) and EDP.

The boron etch-stop method has been successfully used to stop the etching of *n*-Si at a *p*⁺ interface (i.e. a boron doped region). However, there are several disadvantages in the use of a boron etch-stop. First, a high level of boron doping is needed which produces a residual tensile strain in the silicon [3.26]. This limits the use of additional epilayers and prohibits the fabrication of ICs within the layer [3.27].

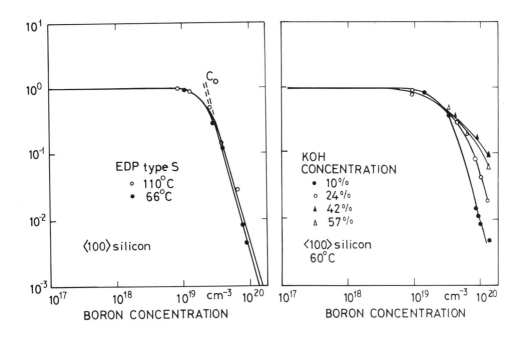

Figure 3.8 Effect of boron doping on etch-rate [3.20].

The limitation of a boron etch-stop can be generally overcome through the use of an Electro-Chemical Etch-stop (ECE). In ECE, the silicon wafer containing a *p-n* junction is immersed in an electrochemical cell. Figure 3.9 shows the schematic layout of a 3-electrode electrochemical etch-stop together with the equivalent electrical circuit using standard notation for the electrode symbols. An anodic contact is made to the *n*-Si and the potential held at a level higher than the passivation potential ~0.6 V [3.28]. The *p*-Si is then electrochemically etched potentiostatically until the *n*-Si layer is reached. At this point the cell current (normally at the level of diode leakage which can be compensated by a fourth electrode if necessary) substantially increases and the etching process is stopped at its peak value. ECE is a useful technique because the doping levels are compatible with IC technology and the process is well controlled to ~1 μm.

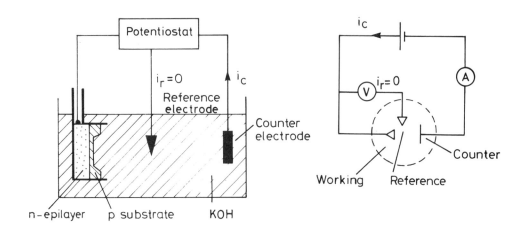

Figure 3.9 Electrochemical etching apparatus. After [3.28] (© 1989 IEEE).

More recently, some novel electrochemical techniques have been developed such as photoelectrochemical etching in which laser light is used to either cause (*p*-Si) or passivate (*n*-Si) the chemical etching process. These have some attraction due to their good control of the etching process.

3.3.5 *Bulk etched structures*

The anisotropic wet etching of silicon can produce a variety of structures that depend upon the mask design, crystal orientation and etch time. Figure 3.10 shows the typical patterns resulting from anisotropic etching of <100> silicon and <110> silicon with the <111> etch-rate set to zero. Similar structures can also be produced using the boron or electrochemical etch-stop as shown in Figure 3.11. Thus, flat bottomed cavities with flat or sloping sidewalls can be made or v-shaped grooves. In addition, the removal of material from both sides of a mask can produce *mesa* or flat-topped structures.

In principle, we would expect to get a perfect pyramid or oblong structure; however corner undercutting may take place due to etching in other planes (e.g. {122}, {141}). The presence of corner undercutting can usually be prevented through the use of either better etchants or compensating masks. The exact modification of the mask is not easy to determine and may require some considerable experimentation before the ideal geometry is found. Some common types of structures that can be isotropically or anisotropically etched in silicon (and other materials) are listed in Table 3.3 below.

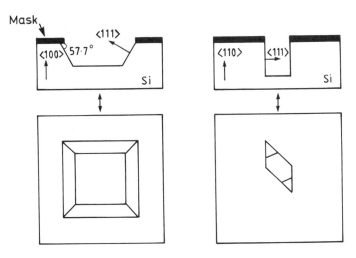

Figure 3.10 Anisotropically etched cavities in silicon wafer.

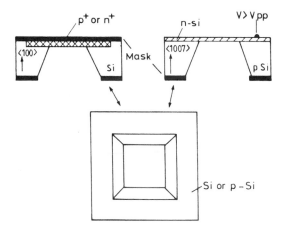

Figure 3.11 Boron-doped or electrochemically etched cavity in silicon wafer.

Table 3.3 Common types of bulk micromachined structures.

Method used	Type
Isotropic etching	Spherical/spherical domes.
	Spherical/spherical cavities.
	Grooves with circular/ellipsoidal cross-sections.
	Cusped pillars.
Anisotropic etching	Angled mesas or dimples.
	Angled and pyramidal cavities.
	Angled apertures and nozzles.
	Diaphragms, cantilevers and bridges.

Figure 3.12 Bulk micromachined silicon pressure sensor. From [3.29].

The first silicon (pressure) sensor was isotropically micromachined by Honeywell in 1962 using a combination of wet etching and dry etching and oxidation processes. One of the earliest anisotropically etched silicon pressure sensors was made by Greenwood in 1984 [3.29] and has led to considerable commercial success for Druck Ltd, UK. Figure 3.12 shows a photograph of this resonator which consists of two coupled oscillators with either a capacitive [3.29] or optical pick-up (not shown). In both cases the resonant frequency of the structures is calibrated as a function of pressure (and temperature) to produce a precision pressure sensor (§7.6.4).

3.4 Surface Micromachining Techniques

Bulk micromachining has been extensively used to fabricate cantilevers, bridges and other mechanical structures in single-crystal silicon through the use of ECE and reactive ion etching techniques (§7.2). The technique has been extended so that microstructures can be made of thin layers of polysilicon, amorphous silicon, silica, oxides or polyimides. This has enabled the micromachining of insulating materials or materials with more favourable physical properties than crystalline silicon. In 1977, the micromachining technique was first applied to materials other than silicon [3.31] and the method essentially involved the use of a sacrificial thin film. This technique is now widely known as surface micromachining [3.12-3.15]. The exact details of the technique depend upon the microsensor required but Figure 3.13 shows the essential steps in the fabrication of a silicon sensor.

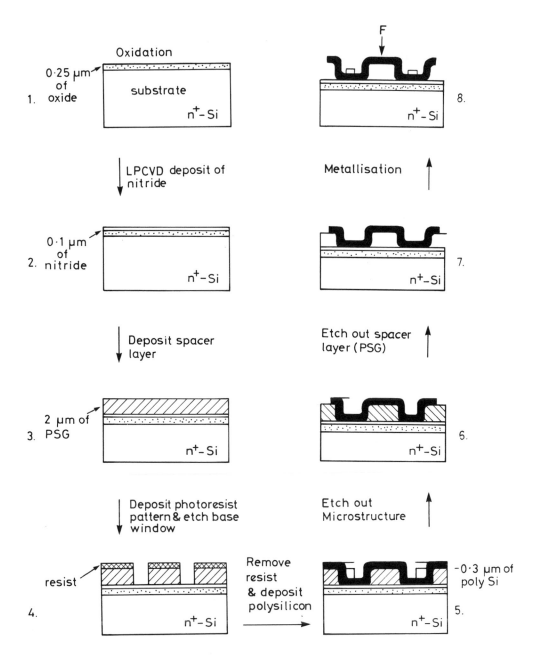

Figure 3.13 Typical fabrication steps required to surface micromachine a structure.

First the wafer is thermally oxidised to give a 0.25 μm layer of SiO$_2$ followed by the LPCVD of a thin layer of silicon nitride ~100 nm thick (Steps 1 and 2). This acts as an etch-stop for KOH or some other etchant.

Then a sacrificial spacer layer such as 2 µm of a phosphosilicate glass is sputtered onto the surface (Step 3). Next, following the deposition of a photoresist layer, UV lithography is used to etch out a base pattern (Step 4). Then the material for the micromechanical structure is deposited, e.g. 0.3 µm of polysilicon and UV lithography is used once again to etch out the microstructure pattern (Steps 5 and 6). Finally, the spacer layer is selectively etched out from underneath the polysilicon layer to leave the free-standing structure (Step 7) and electrical contacts are produced lithographically (Step 8).

In this example we have produced a doubly-supported cantilever beam that could be used as a capacitive force or pressure sensor with, for example, the n^+-substrate acting as a gate. A recent discussion of surface micromachining has been made by Howe [3.32].

The development of the surface micromachining technique has led to the fabrication of a wide variety of micromechanical structures that can be used as microsensors or microactuators (Chapter 7). Figure 3.14 shows two examples of surface micromachined structures, a microresonant accelerometer (§7.5) and a four pole motor [3.33].

In addition, bulk and surface micromachining processes can be combined to make structures such as sub-micron particle filters [3.34] or simply produce deep cavities. Most importantly, it is now possible to integrate IC circuitry with the micromachined structure which is often necessary when one considers the small capacitances (~fF) of such sensors. An example of this is a precision

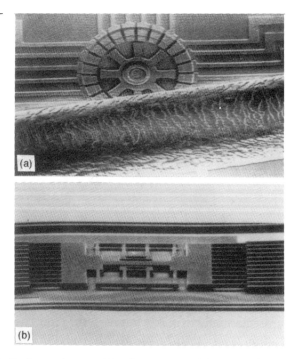

Figure 3.14 Surface micromachined structures: (a) a microflexure resonator for use as an accelerometer microsensor and (b) a micromotor [3.33].

microaccelerometer which was based upon a resonant structure and associated CMOS circuitry [3.35]. It achieved a very high sensitivity in the range of a few µg. A more discussion of integrated microsensors is found in later chapters on smart sensors (11) and microsensor array devices (12).

3.5 Silicon Bonding Techniques

There is often the need to bond a silicon microstructure or substrate onto another surface, perhaps of a similar or dissimilar material.   For example, in the manufacture of the bulk micromachined pressure sensor discussed above [3.29], we need to bond a glass substrate onto the etched silicon to form a hermetically-sealed cavity.   There are four main types of bonding technology and these are listed below in Table 3.4 with some details.

Table 3.4 Silicon bonding technologies.

Technology	Bonding temperature	Adhesion with
Polymer	Low (130 - 350°C)	Resists, polyimides etc.
Fusion	Wide range (100 - 600°C)	Glass frits, silicates, phosphate etc.
Anodic	Medium (400 - 800°C)	Soda glasses.
Reaction	Medium (≥300°C)	Directly to silicon.

The exact choice of bonding technology depends upon the particular application requirements.   For example, a layer of negative photoresist or polyimide can be used in polymer bonding at relatively low temperatures. However, fusion bonding is more popular in which a silicate or phosphate can be used as an intermediate layer to join surfaces at temperatures between 100°C and 500°C.   An alternative to silicates is the use of a glass frit containing a metal additive (e.g. Pb) which fuses the surfaces together at about 500°C.   The intermediate layer can be deposited in a variety of ways such as spin-coating, spraying or CVD.

Another type of bonding technology used is anodic bonding of silicon to a soda glass (i.e. a glass with a high impurity level). At moderate temperatures (~ 500°C) the sodium ions become mobile and, field-assisted, drift towards the interface. The resultant electrostatic field pulls the glass and silicon surface close together so that oxygen ions can move and form new Si-O bonds at the interface. The disadvantage of this technique is the need for a high working temperature and high voltage. A voltage in the kV range (temperature-dependent) will fuse the materials but may lead to damage in any associated IC circuitry.

Finally, reaction bonding of silicon to silicon is an attractive technology in some applications [3.36]. The process takes place at room temperature, requires no glass frit or other intermediate layer, and does not require a high driving voltage. In reaction bonding, the two surfaces form a hydrogen bridge when pushed together. Annealing at temperatures above 300°C breaks the O-H bonds to form water. The H_2O then dissociates to form hydrogen which diffuses out through the silicon, and oxygen which bonds directly onto any Si dangling bonds. The two processes of bond breaking and water reacting may be written as:

$$1.\ Si\text{-}OH + HO\text{-}Si \rightarrow Si\text{-}O\text{-}Si + H_2O$$
$$2.\ Si + H_2O \rightarrow H_2 + Si\text{-}O$$

(3.10)

There are now many types of bonding technologies used in sensor fabrication: some are required in the fabrication of the sensing device and others for mechanical support or thermal grounding. Some other types of bonding that include metallic material bonding (as mentioned above in IC wire-bonding) are:

- eutectic bonding;
- compression and thermocompression metallic bonding;
- ultrasonic welding;
- laser welding.

Readers particularly interested in microsensor bonding techniques are referred to references [3.37]-[3.39].

In the next chapter we consider dry etching and other more specialised methods of micromachining materials for microsensors. A discussion is also made of the processing of active materials and their key physical properties.

Suggested further reading

Readers are referred to the following texts that provide some background information for this chapter on conventional silicon processing:

Sze SM: *Semiconductor Devices: Physics and Technology* (1985). Published by Wiley & Sons Inc., New York. ISBN 0-471-09837-X. (523 pages. Details of silicon process technology)

Fung CD, Cheung PW, Ko WH and Fleming DC (eds.): *Micromachining and Micropackaging of Transducers* (1985). Published by Elsevier, Amsterdam. ISBN 0-444-42560-8. (244 pages. Papers on selected topics in this field)

References

3.1 Hoerni JA (1960) Planar silicon transistors and diodes. *IRE Electron. Device Mtg., Washington, DC., USA.*

3.2 Verhofstadt PWJ (1976) Evaluation of technology options for LSI processing elements. *Proc. IEEE*, **64**, 842-851.

3.3 Sze SM (1985) *Semiconductor Devices: Physics and Technology*, John Wiley & Sons, New York, 523 pp.

3.4 Sze SM (ed.) (1988) *VLSI Technology*, McGrawHill, New York.

3.5 Casey HC and Pearson GL (1975) Diffusion in semiconductors, in JH Crawford and LM Slifkin (eds.) *Point Defects in Solids*, **Vol. 2**, Plenum Press, New York.

3.6 Crank J (1975) *Mathematics of Diffusion*, Oxford University Press, Oxford, UK, 414 pp.

3.7   Fair RB (1981) Concentration profiles of diffused dopants, in F.F. Wang (ed.) *Impurity Doping Processes in Silicon*, North-Holland, Amsterdam.

3.8   Pickar KA (1975) Ion implantation in silicon, in R. Wolfe (ed.) *Applied Solid State Science*, **Vol. 5**, Academic Press, New York.

3.9   Senturia SD and Smith RL (1988) Microsensor packaging and system partitioning. *Sensors and Actuators*, **13**, 221-234.

3.10 Smith RL and Collins SD (1988) Micromachined packages for chemical microsensors. *IEEE Trans. Electron Devices*, **35**, 787-792.

3.11 Kaminsky G (1985) Micromachining of silicon mechanical structures. *J. Vac. Sci. Technol. B*, **3**, 1015-1024.

3.12 Robbins H and Schwartz B (1959) Chemical etching of silicon; Part I. *J. Electrochem. Soc.*, **106**, 505-508.

3.13 Robbins H and Schwartz B (1960) Chemical etching of silicon; Part II. *J. Electrochem. Soc.*, **107**, 108-111.

3.14 Schwartz B and Robbins H (1961) Chemical etching of silicon; Part III. *J. Electrochem. Soc.*, **108**, 365-372.

3.15 Schwartz B and Robbins H (1976) Chemical etching of silicon; Part IV. *J. Electrochem. Soc.*, **123**, 1903-1909.

3.16 Farooqui M (1992) Wet etching of silicon. *Course notes on Micromachining of Materials, University of Southampton, March 1992.*

3.17 Theunissen MJ, Appels JA and Verkuylen WHC (1970) Application of electrochemical etching of silicon to semiconductor device technology. *J. Electrochem. Soc.*, **117**, 959-965.

3.18 Tenney AS and Ghezzo M (1973) Etch rates of doped oxides in solutions of buffered HF. *J. Electrochem. Soc. Solid-state Sci. Technol.*, **120**, 1092-1095.

3.19 van Gelder W and Hauser VE (1967) The etching of silicon nitride in phosphoric acid with silicon dioxide as a mask. *J. Electrochem. Soc. Solid-state Sci. Technol.*, **120**, 869-871.

3.20 Siedel H (1989) The mechanism of electrochemical and anisotropic silicon etching and its applications. *Proc. 3rd Toyota Conf. on Integrated Micromation Systems, Nissin, Aichi, Japan, October 1989.*

3.21 Siedel H, Csepregi L, Heuberger A and Baumgartel H (1990) Anisotropic etching of crystalline silicon in alkaline solutions, I and II. *J. Electrochem. Soc.*, **137**, 3612-3632.

3.22 Mehregany M and Senturia SD (1988) Anisotropic etching of silicon. *Sensors and Actuators*, **13**, 375-395.

3.23 Schnakenberg U, Benecke W and Lochel B (1990) NH_4OH-based etchants for silicon micromachining. *Sensors and Actuators A*, **21-23**, 1031-1035.

3.24 Bean KE (1978) Anisotropic etching of silicon. *IEEE Trans. Electron. Devices*, **25**, 1185-1193.

3.25 Reisman A, Berkenblit M, Chan SA, Kaufman FB and Green DC (1979) The controlled etching of silicon in catalysed ethylenediamine-pyrocatechol-water solutions. *J. Electrochem. Soc.*, **126**, 1406-1415.

3.26 Siedel H and Csepregi L (1988) Advanced methods for the micromachining of sensors. *7th Sensor Symposium, IEE of Japan, Tokyo, May 1988*, 1-6.

3.27 Wise KD (1985) Silicon micromachining and its application to high performance integrated sensors, in *Micromachining and Micropackaging of Transducers* (eds. C.D. Fung, P.W. Cheung, W.H. Ko and D.G. Fleming), Elsevier, Amsterdam, pp. 3-18.

3.28 Kloeck B, Collins SD, de Rooij NF and Smith RL (1989) Study of electrochemical etch-stop for high precision thickness control of silicon membranes. *IEEE Trans. on Electron Devices*, **36**, 663-669.

3.29 Greenwood JC (1984) Etched silicon vibrating sensor. *J. Phys. E: Sci. Instrum.*, **17**, 650-652.

3.30 Pitcher RJ, Foulds KW, Clements JA and Naden JM (1990) Optothermal drive of silicon resonators: the influence of surface coatings. *Sensors and Actuators A*, **21**, 387-390.

3.31 Petersen KE (1977) Micromechanical light modulator array fabricated on silicon. *Appl. Phys. Lett.*, **31**, 521-523.

3.32 Howe RT (1988) Surface micromachining for microsensors and microactuators. *J. Vac. Sci. Technol.*, **B6**, 1809-1813.

3.33 Muller RS (1990) Microdynamics. *Sensors and Actuators A*, **21-23**, 1-8.

3.34 Kittilsland G, Stemme G and Norden B (1990) A submicron particle filter in silicon. *Sensors and Actuators A*, **21-23**, 904-907.

3.35 Rudolf F, Jornod A, Bergquist J and Leuthold H (1990) Precision microaccelerometers with μg resolution. *Sensors and Actuators A*, **22**, 297-302.

3.36 van den Vlekkert HH, Decroux M and de Rooij NF (1987) Glass encapsulation of chemical solid-state sensors based on anodic bonding. *Proc. Transducers '87, Tokyo*, pp. 730-733, 1987.

3.37 Ko WH, Suminto JT and Yeh GH (1985) Bonding techniques for microsensors, in *Micromachining and Micropackaging of Transducers* (eds. C.D. Fung, P.W. Cheung, W.H. Ko and D.C. Fleming), Elsevier, Amsterdam, pp. 41-62.

3.38 Reichl H (1991) Packaging and interconnection of sensors. *Sensors and Actuators A*, **25-27**, 63-71.

3.39 Harendt C, Hofflinger B, Graf H-G and Penteker E (1991) Silicon direct bonding for sensor applications: characterisation of bond quality. *Sensors and Actuators A*, **25-27**, 87-92.

Problems

3.1 Briefly describe the processing steps that are involved in the wet etching of a cavity in single-crystal silicon.

3.2 Sketch out the ideal shapes that would be etched out of single-crysal silicon using square, triangular and cross-shaped masks after various times.

3.3 Equation (3.1) describes the process of a dopant diffusing into a semiconductor material. Solve the equation for a semi-infinite planar structure and a thin plate structure. Sketch the profiles at various times. Would you expect the same profile in practice, e.g. for arsenic doping of crystalline silicon?

3.4 State the etchants that could be used to: (a) anisotropically etch silicon, (b) isotropically etch silicon and (c) etch silicon nitride. What factors influence the etch-rate?

4. Specialised Materials Processing

Objectives

- [ ] To discuss materials processing for microsensor fabrication

- [ ] To describe specialised dry micromachining techniques

- [ ] To present the properties of sensor materials

- [ ] To summarise the performance of micromachining technologies

4.1 Introduction

The wet etching techniques that have been described in the last chapter are fundamental to the materials processing of microsensor fabrication. Yet there are other types of micromachining techniques that can have a significant technological advantage over wet etching. These are generally known as *dry etching* techniques because the technique does not require the immersion of the workpiece in a solution in order to carry out the micromachining process.

There are three principal types of dry etching techniques:

1. Reactive etching (including chemically assisted);
2. Ion etching (physical sputtering and milling);
3. Focused energy beam techniques (ions, photons, electrons).

Reactive etching consists of the use of either a reactive plasma (usually radio frequency or electron cyclotron resonance induced) or a reactive ion to remove the material. Advantages of using dry etching techniques in general over wet etching include the fact that the etching process can be more precisely controlled and in focused ion etching it is possible to image the workpiece. However, a disadvantage is that dry etching reactors always require a vacuum system and relatively sophisticated instrumentation. So far all of the micromachining techniques that we have considered remove material uniformly over a wide area. *Ion etching* or showered ion beam milling is another such distributed process which physically sputters material off a surface by bombarding it with ions. In contrast, *focused energy beam techniques* offer the possibility of the direct machining of small structures by the localised bombardment of ions, photons or electrons. We shall now consider each of these three types of dry etching techniques in turn, after which a consideration is made of the properties of the materials used in microsensors. This is followed by a description of some specialised deposition techniques for ultrathin, thin and thick films. Finally, a comparison is made between all the micromachining techniques discussed so far.

4.2 Reactive Ion Etching

Plasmas were first used in the 1970s to etch materials and considerable advances have been made in the technology since then [4.1, 4.2]. Plasma fields are generated in a glass tube or barrel with the Radio Frequency (R.F.) power coupled either inductively or capacitively into the gas, e.g. oxygen, at low pressures. Oxygen has been used for the isotropic etching of resists in which the excited oxygen atoms react with the organic resists to produce water and CO or CO_2. Figure 4.1(a) shows a typical barrel-type reactor in which a coil wrapped around the barrel inductively couples the R.F. power into the gas. Electrons are accelerated in the

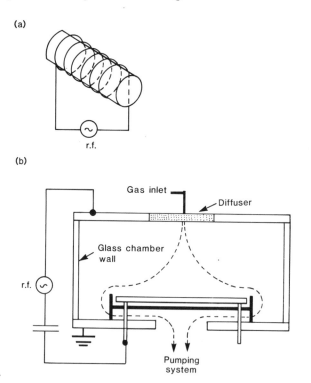

Figure 4.1 A schematic diagram of (a) barrel-type R.F. inductively coupled and (b) parallel-plate ion etching reactors.

field and collide with the gas molecule to produce an excited gaseous species. For example, oxygen reacts to produce excited species $[O^*]$ such as O^- in the following reaction:

$$O_2 + [e^-] \rightarrow 2O^- + 3[e^-] \qquad\qquad (4.1)$$

At a pressure of 1.5 Pa, a voltage of -440 V and a power of 30 W the oxygen plasma etches the polymers PMMA and poly(styrene) at 0.40 μm/min and 0.13 μm/min, respectively [4.3].

Figure 4.1(b) shows a schematic diagram of a parallel-plate reactive ion etching reactor in which the field is capacitively rather than inductively coupled. This arrangement allows the anisotropic dry etching of metals and semiconducting (inorganic) materials as well as organic ones. The anisotropic etching is caused by a large potential difference (or so-called self-bias) which appears across the electrodes (about 100 to 800 V) and influences the chemical reaction.

Figure 4.2 illustrates the application of reactive ion etching in VLSI and microsensor systems. Figure 4.2(a) shows an electron micrograph of an array of 0.3 μm GaAs/GaAlAs diodes fabricated using an alkane plasma (C_2H_6) reactive ion etch [4.4]. The resolution of the etch process is clearly very high as indicated by the good definition visible. Figure 4.2(b) shows a mesa structure etched out of InP using a CH_3Cl plasma etch at 300 W power for 25 min [4.5].

More recently, microwave electron cyclotron resonance reactors have been developed to improve the reactive-etching processes. In this technique the reactive ions are freed from the plasma to enable delicate IC or other work. The standard reactors operate at frequency of 2.45 GHz with a magnetic field of below 88 mT. The range of reactive-ion gases used and materials etched are summarised in Table 4.1 below, together with some comments on the nature of the source gas used.

Table 4.1 Source gases and some of the materials suitable for reactive etching.

Source gas	Materials etched	Comments
O_2	Photoresist (PMMA and polystyrene)	Explosive hazard
CF_4 (Freon)	Si and III-V compounds	Toxic gas
Cl_2	Polysilicon, SiO_2	Corrosive gas
Alkanes (e.g. C_2H_6)	III-V and II-VI compounds	Flammable gas
$SiCl_4$, SiF_4	III-V compounds and SiO_2	Toxic gas
CF_3H	Silicon nitride	Toxic gas

The resolution of reactive ion etching is clearly of relevance to both VLSI and microsensor fabrication as its diffraction-limited resolution of less than 0.1 μm exceeds that of conventional UV lithography. It can also be used to machine a wide variety of materials. However, the advantage of reactive ion etching over

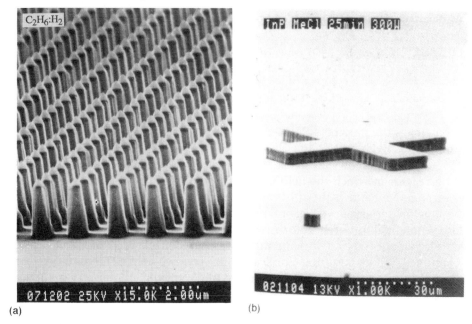

(a)   (b)

Figure 4.2   Photographs of (a) 0.3 μm GaAs/GaAlAs diodes and (b) a mesa microstructure in InP, formed by reactive ion etching techniques. From [4.4, 4.5].

current lithography is not as significant when compared with X-ray lithography or electron beam lithography which both have a higher spatial resolution than UV.

4.3 Ion etching

The process of ion etching involves the bombardment of a workpiece with ions to remove material by physical rather than chemical means. The sputtering process does not heat the workpiece to its melting point as in the cases of laser or microelectrical discharge micromachining (see later) which can be a distinct advantage.   Originally, ion etching was regarded as too slow for practical implementation but the recent development of the Kaufman high current density source has enabled etch-rates of the order of μm/min over wafer sized areas. The Kaufman ion source uses electron bombardment to ionise the gas (usually argon), then accelerates and channels the ions into a uniform beam by a set of graphite grids [4.6].   Kaufman ion sources offer many advantages for micromachining application including a high, wide and uniform current density with a reasonable degree of collimation. Figure 4.3 shows the schematic arrangement of a reactor developed at Warwick University for the showered ion beam milling of materials [4.7]. The reactor may be operated manually or linked via an Intelligent Crate Controller (ICC) to a PC. The reactor produces an ion source of energy up to 1.5 keV and a current density of 25 mA/cm^2 covering a beam diameter of 3 cm.

The ion beam milling of a material depends upon many parameters such as the type of target material, ion species, ion energy, angle of incidence etc. The etch-rate E in nm per minute per cm^2 per milliamp of beam current can be calculated approximately from the sputter yield rate Y as follows:

$$E(\theta) = k \frac{Y(\theta)\cos(\theta)}{N_Z} \tag{4.2}$$

where θ is the incident angle between the beam and the vector normal to the

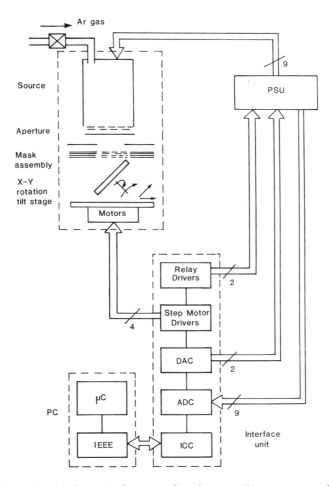

Figure 4.3 A schematic diagram of ion beam milling apparatus [4.7].

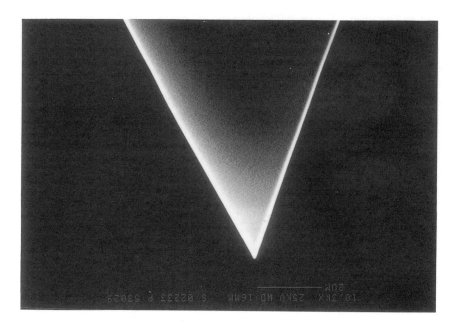

Figure 4.4 Micrograph of a diamond tip sharpened to a 10-15 nm diameter by showered ion beam etching [4.9].

surface, k is a constant of value 9.6×10^{24} and N_z is the atomic density of the target (atoms/cm^3).   The maximum etch rate (various angles) in nm/min/mA is typically 100 for Au, 75 for Al, 150 for GaAs, 80 for AZ1350 photoresists and 170 for COP electron resist [4.8].

Showered ion etching has a wide range of micromachining applications. One important application is its use to profile styli for use in mechanical microprobes. Figure 4.4 shows a micrograph of an ion etched diamond stylus to an estimated radius of between 10 and 15 nm [4.9]. The technique could also be used to texture the surface of optical and solar cells, and patterning in IC or microsensor device fabrication (although it is difficult to make ion-resistant masks). Table 4.2 shows the ion etch-rate for various materials [4.10]

Table 4.2   Ion etch-rates of various materials at an ion energy of 500 eV [4.10].

Material	Etch-rate (nm/min)	Material	Etch-rate (nm/min)
Aluminium	30-70	Silicon	22-50
Gold	105-150	GaAs	65
SiO$_2$	28-42	Resist AZ1350	20-42
Al$_2$O$_3$	83	Riston 14	25

A second type of ion etching is called Focused Ion Beam Milling (FIBM). This technique uses a liquid metal source (e.g. gallium) rather than argon and focuses the ion beam down to a submicron diameter. The liquid metal ion source can employ a capillary feed (or needle) system and an acceleration voltage of typically 5 to 50 keV. Beam spots down to 0.1 μm diameter are possible which are limited by the chromatic aberration caused by the energy distribution within the ion beam. Existing applications include the repair of X-ray masks [4.11] and VLSI circuits [4.12] and the technique offers enormous potential for the fabrication of novel microsensors.

4.4 Laser Micromachining

Lasers have been used in materials processing for the past 20 years or so. Initially, the carbon dioxide (CO_2) laser and neodymium yttrium aluminium garnet (Nd:YAG) were favoured over other traditional laser types such as the He:Ne or dye laser because of their higher power output.

CO$_2$ lasers use a mixture of helium (83%), nitrogen (16%) and carbon dioxide (6%) as the lasing material. The CO_2 laser produces a collimated coherent beam in the infrared region of wavelength (10.6 μm) characteristic of the active material (i.e. CO_2). The wavelength is strongly absorbed by glasses or polymers and so either mirrors or ZnSe lenses are used to handle the beam. The use of axial or cross flow of the gas mixture permits the generation of continuous beam powers of up to 25 kW.

The Nd:YAG laser produces a collimated coherent beam in the near-infrared region of wavelength 1.06 μm and is run in a pulsed mode, and peak power pulses of up to 20 kW are now available. The shorter wavelength allows the use of optical glasses to control the beam path. A typical unfocused beam may have a diameter of 10 to 40 mm and a power density of ~ 1 W/mm^2.

In 1975 the excited dimer (excimer) laser was invented in which a diatomic molecule such as N_2 or H_2 was used as the lasing material. Since then rare gas halide lasers have become more common but are still referred to as excimer lasers. Excimer lasers cover a range of wavelength in the ultraviolet range from 157 nm (F_2) to 353 nm (XeF) with 193 nm (ArF), 248 nm (KrF) and 308 nm (XeU) being particularly useful intermediate wavelengths. Peak powers typically range from 3 MW (F_2) to 50 MW (KrF or XeU) in pulses lasting a few tens of nanoseconds.

There are various applications of lasers in micromachining; some of these are:

- dry etching of polymers;
- exposure of resist masks;
- laser induced chemical etching of semiconductors, metals, ceramics;
- focused beam milling of plastics, glasses, ceramics, metals.

The excimer laser has several distinct advantages in the dry etching of polymers over the CO_2 and Nd:YAG lasers. For instance, the wavelengths are more compatible with the chemical bond energies in organic compounds and tend to produce less thermal damage. Figure 4.5 shows a micrograph of a polycarbonate film micromachined by a KrF excimer laser at 248 nm and 2.5 J/cm^2 and a suitable mesh pattern to produce the honeycomb structure [4.13]. Similar patterns can be micromachined in polyimides with a typical laser etch-rate of about 0.5 μm per pulse of 1 J/cm^2.

Excimer lasers can also be used as more efficient UV sources for photolithographic exposure of resists for VLSI circuits. In addition excimers are used to assist the wet chemical etching of VLSI circuits. The process is called laser-induced chemical etching in which, for example, chlorine is used as the etchant and is photo-dissociated to react with the semiconductor [4.14]. An attraction of this technique is that it produces practically no debris at the edges and thus has an excellent spatial resolution.

The other major application of lasers is the <u>focused</u> beam milling of plastics, glasses, metals and ceramics etc. The minimum diameter of the focused beam, d_{min}, is normally determined from the diffraction limits of the imaging system. So the theoretical resolution is approximately given by,

$$d_{min} \approx \frac{k\lambda}{N_a} \approx \frac{2k\lambda f}{d_o} \qquad (4.3)$$

where k is a dimensionless constant (usually having a value of 0.6 to 0.8), λ is the wavelength of the radiation, N_a the numerical aperture ($\sim d_o/2f$), d_o is the diameter of the beam at the lens and f is the focal length of the lens. Thus, the principal way

Figure 4.5 Micrograph of a polycarbonate film micromachined using a KrF eximer laser and mask [4.13]. (Courtesy of Exitech Ltd, UK.)

of increasing the resolution (while having a reasonable focal depth) is by reducing the wavelength.

Table 4.3 lists some of the characteristics of the CO_2, Nd:YAG and excimer (KrF) laser relating to micromachining purposes (FIBM and EDM are added for comparison).

Table 4.3 Some characteristics of laser micromachining systems.

System	Feature resolution (μm)	Focal depth (μm)	Wavelength (μm)	Average power (μW)
CO_2	17	60	10.6	1,000 to 25,000
Nd:YAG	1.7	6.0	1.06	1,000 to 2,000
KrF	0.4	1.4	0.248	≤ 150
FIBM	0.05	10 to 100	~0.002	Medium
EDM	>50	n/a	n/a	Low

Lasers have been successfully used to drill ultra-small holes in a variety of materials. Focused beams from both Nd:YAG and excimer lasers can drill holes in a variety of materials in the range 30 to 50 μm.

Figure 4.6 shows 50 μm holes drilled through a single human hair by an excimer laser (Exitech Ltd, UK). Microholes, grooves and other structures can be micromachined in materials such as alumina, silicon nitride and zirconia. The smallest feature for excimer lasers is about 0.8 μm in thin metal foils. It is highly likely that excimer laser micromachining will soon be a major process in the

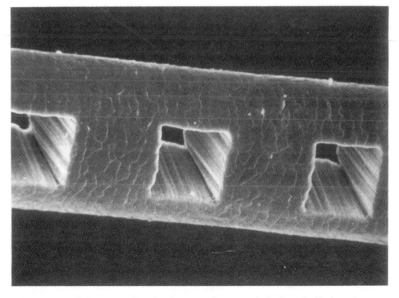

Figure 4.6 Micrograph of a human hair with holes drilled in by an excimer laser [4.13]. (Courtesy of Exitech Ltd, UK.)

micromachining of microsensors. The requirement for even smaller linear features necessitates the use of a FIBM technique.

For the sake of completeness, Table 4.3 includes the performance of a completely different technology called Electro-Discharge Machining (EDM). EDM is traditionally used to drill holes for punch and dies, but it has a limited application in micromachining. The principle of EDM is that a spark is generated by a high voltage between a wire electrode and the conducting workpiece. This causes the workpiece (and capillary-fed wire electrode) to erode away leaving a hole. The technique has been used to make fine nozzles in metals for ink jet printers but its limiting feature size of ~50 μm is determined by the diameter of the wire. The main advantages of EDM are that it can mill through metals at much higher rates using a simpler technology.

4.5 Materials for Microsensors

4.5.1 Passive materials

We have just described how a variety of materials can be processed in conventional IC microelectronic technology, bulk micromachining and surface silicon micromachining technologies. This satisfies our need in microsensors to process bulk crystalline semiconducting materials, such as Si and GaAs, as well as thin films of polysilicon, amorphous silicon, oxides (e.g. silica), polyimides, and nitrides (e.g. Si_3N_4). In addition there is a requirement to make electrical connections to the microsensor which demands the processing of thin metal or alloy films such as Al, Au, Al-Cu-Cr, Ti and Mo (or heavily doped silicon or polysilicon). All of these materials may be described as *passive* materials because they are generally used to provide either a mechanical structure or electrical connection.

Tables 4.4 and 4.5 summarise some of the basic physical properties of the passive materials that determine their applicability to a particular sensing application. Some of these passive materials can also be used as an active material, e.g. silicon, and so will also appear in the table listing common active materials in the next section.

When designing microsensors, a material may be required to have a low thermal or electrical conductivity to insulate one component from another, have a thermal conductivity similar to an active material, or alternatively it may require good mechanical properties. Stresses induced by materials with different thermal properties can often be a problem with silicon microsensors and so a careful choice of substrate material is essential. But it should be remembered that the film property can vary with deposition technique or purity. Thus the values shown in Tables 4.4 and 4.5 should only be used as a guide when precise figures are not known. Prinicpal references to the source material are shown in the tables.

However, it is also necessary to deposit and use active materials in microsensors. These materials are essential to the sensing process employed in various types of microsensors, such as photosensitive (§6.3), piezoelectric (§7.7.2), magnetoresistive (§8.3), and chemoresistive (§9.2) films.

Table 4.4   Physical properties of common non-metallic passive materials used in microsensor technology [4.15, 4.16]. Values are taken at 293 K where appropriate.

Material: Property	Si (c)	GaAs (c)	SiO$_2$ (quartz)	Si$_3$N$_4$ (c)
Density, ρ_m (kg/m^3)	2,330	5,316	1,544	3,440
Melting point, T_{mp} (°C)	1,410	1,510	1,880	1,900
Boiling point, T_{bp} (°C)	2,480	-	2,500	-
Thermal conductivity, κ (W/m/K)	168	47	6.5, 11	19
Specific heat capacity, c_p (J/K/kg)	678	350	730	-
Temperature expansivity, α_l (10^{-6} /K)	2.6	5.7	7, 12	0.8
Dielectric constant, ε_r	11.7	12	4.5, 4.3	7.5
Young's modulus, E (GPa)	190	-	380	380
Yield strength, Y (GPa)	6.9	-	14	14

4.5.2 Active materials

It is often necessary to carry out the processing of additional materials in order to fabricate the sensor, although in some cases the sensor may be simply a temperature-sensitive *n-p* junction (§5.5) or a photosensitive silicon transistor (§6.3). There is a large range of additional materials currently used in microsensors and these often take the form of thin or thick films and play an active role in the sensing system. Some of these active materials can be deposited using IC-compatible deposition techniques such as CVD or LPCVD but others need special techniques such as electrochemical deposition as in the case of conducting polymers. Moreover, these materials may place special demands upon the materials processing as they may not be able to withstand subsequent processing due to a low thermal stability or a problem with contamination. In the next section we discuss some special deposition techniques for active ultra-thin (less than 100 nm), thin (100 nm to 1 μm) and thick (1 to 100 μm) active films.

In Table 4.6, the properties of a variety of active materials are given along with the type of sensor and manner in which the film can be deposited. For example, ZnO is commonly used as the active material in piezoelectric sensors (§7.7, §9.7). It can be sputtered or chemically vapour deposited onto suitable substrates.

Table 4.5   Physical properties of some common metallic passive materials used in microsensor technology [4.16].

Material: Property:	Al	Au	Cr	Ti	W
Density, ρ_m (kg/m^3)	2,699	19,320	7,194	4,508	19,254
Melting point, T_{mp} (°C)	660	1,064	1,875	1,660	3,422
Boiling point, T_{bp} (°C)	2,467	2,967	2,482	3,313	5,727
Electrical conductivity, σ (10^3 S/cm)	377	488	79	26	183
Temperature coefficient of resistance, α_r (10^{-4} /K)	43	34	30	38	-
Work function, ϕ (eV)	4.3	5.1	4.5	4.3	4.6
Thermal conductivity, κ (W/m/K)	236	319	97	22	177
Specific heat capacity, c_p (J/K/kg)	904	129	448	522	134
Linear expansivity, α_l (10^{-6} K^{-1})	23	14	4.9	8.6	4.5
Young's modulus, E (GPa)	70	78	279	~40	411
Yield strength, Y (MPa)	50	200	-	480	~750
Poisson's ratio, ν	0.35	0.44	0.21	0.36	0.28

In general, inorganic semiconducting materials can be processed without too much difficulty due to their good thermal stability. Nevertheless some microsensors will only operate under certain conditions which in turn can place a considerable demand upon the material properties; for instance a pellistor gas sensor only detects methane detection at a high operating temperature (§9.6.2) while a quartz piezoelectric pressure sensor (§7.6.4) requires a high input voltage to achieve a reasonable sensitivity.

4.5.3 *Langmuir-Blodgett deposition of ultra-thin films*

The Langmuir-Blodgett technique was invented by Irving Langmuir and Katharine Blodgett and allows the controlled deposition of monomolecular layers. The ultra-thin film deposition technique is limited to materials that consist of amphiphilic long chain molecules with a hydrophobic molecule at one end and a hydrophilic molecule at the other end [4.17].

Table 4.6 Some important properties of active materials used in microsensors [4.16].

Thermal	ρ_m (kg/m^3)	T_{mp} (°C)	σ (S/cm)	κ (W/m/K)	α_1 (10^{-6}/K)	E_g (eV)
Pt	21,447	1,769	9×10^4	72	8.8	n/a
CdS	4,820	1,750	-	-	-	2.42
PbS	7,500	1,114	-	3	-	0.37
Radiation	ρ_m (kg/m^3)	T_{mp} (°C)	σ (S/cm)	κ (W/m/K)	α_1 (10^{-6}/K)	E_g (eV)
Si	2,330	1,410	4×10^{-2}	168	2.6	1.11
Ge	5,323	937	3×10^{-4}	67	5.7	0.67
GaAs	5,316	1,510	10^{-8}	47	5.7	1.35
Mechanical	ρ_m (kg/m^3)	T_{mp} (°C)	Velocity (m/s)	Delay (ppm/°C)	α_1 (10^{-6}/K)	K^2 (%)
Quartz (AT-cut)	1,544	1,880	5,100	2.8	0.8	1.43
Quartz (ST-cut)	1,544	1,880	4,990	33	-	1.89
LiNbO$_3$ (x-axis)	-	-	4,802	59	-	16.7
Magnetic	ρ_m (kg/m^3)	T_{mp} (°C)	σ (S/cm)	c_p (J/K/kg)	T_c (K)	μ_r (10^3)
Fe (pure)	7,874	1,535	10^5	449	1,043	1,500
NiFe alloy (50:50)	8,200	-	3×10^4	~400	798	75
CoFe alloy (50:50)	8,150	-	2×10^4	~400	1,253	7
Chemical	ρ_m (kg/m^3)	T_{mp} (°C)	σ (S/cm)	Sensitivity (ppm)	α_r (/K)	E_g (eV)
SnO$_2$ (c)	6,950	1,360	Low	1 - 1,000	-ve	3.45
PbPc (c)	1,950	~600	Very low	> 0.001	-ve	~0.7
Poly(pyrrole)	~1,500	~200	10^{-4}-10^{+2}	0.1 - 1,000	-ve	Small

Figure 4.7 shows a general view of a Langmuir-Blodgett trough (NIMA Technology, UK). The Langmuir-Blodgett material floats on the surface of the water and spreads out to form a monomolecular layer. The Langmuir-Blodgett film is transferred from the water by the controlled dipping of the substrate into the water at a constant speed and film surface tension. In this way, films can be built up in thickness, monolayer by monolayer to form up to 100 layers. Biological materials such as steric acid have been deposited using this technique and have been shown (§9.7, §12.5) to produce films sensitive to odours [4.18]. More recently, the technique has been used to deposit phthalocyanine films which are sensitive to oxidising gases such as NO$_2$; this use may have important applications in the field of chemical sensing as well as other areas [4.19, 4.20]. The current difficulties in using Langmuir-Blodgett films are firstly ensuring that the film is continuous and defect-free and secondly overcoming the problem of their low thermal stability.

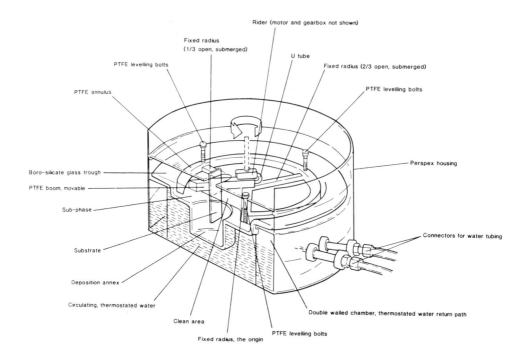

Figure 4.7 General view of a Langmuir-Blodgett trough for depositing ultra-thin films [4.18]. (Courtesy of NIMA Technology, UK.)

4.5.4 Electrochemical deposition of organic polymers

Recent developments in molecular engineering have led to a new range of electropolymerised thin films called *conducting polymers.* Conducting polymers are attractive as passive and active materials because they can be readily deposited and offer a wide range of physical properties. Conducting polymers are usually grown electrochemically in a three electrode cell by either cyclic voltammetry or a potential step method (see [4.21] for the fundamentals of electrochemistry). For example, Figure 4.8 shows the chemical reaction scheme for the polymerisation of pyrrole [4.22]. Pyrrole is a heterocyclic monomer with a conjugated π-system. Poly(pyrrole) can be grown from a range of solvents and with different counter-ions which determine its physical properties. A perchlorate counter-ion leads to a high electrical conductivity of 60 to 200 S/cm while a fluorosulphonate counter-ion leads to a low value of about 0.01 S/cm [4.23]. The choice of the monomer also affects the physical properties of the polymer such as the electrical conductivity, thermal conductivity etc. (see [4.24] for a review of conducting polymers). Although there has been considerable interest shown in the use of conducting polymers as electronic devices, e.g. organic diodes or transistors, there has been less interest until recently in exploiting their possible use as the active material in a gas sensor (§9.2).

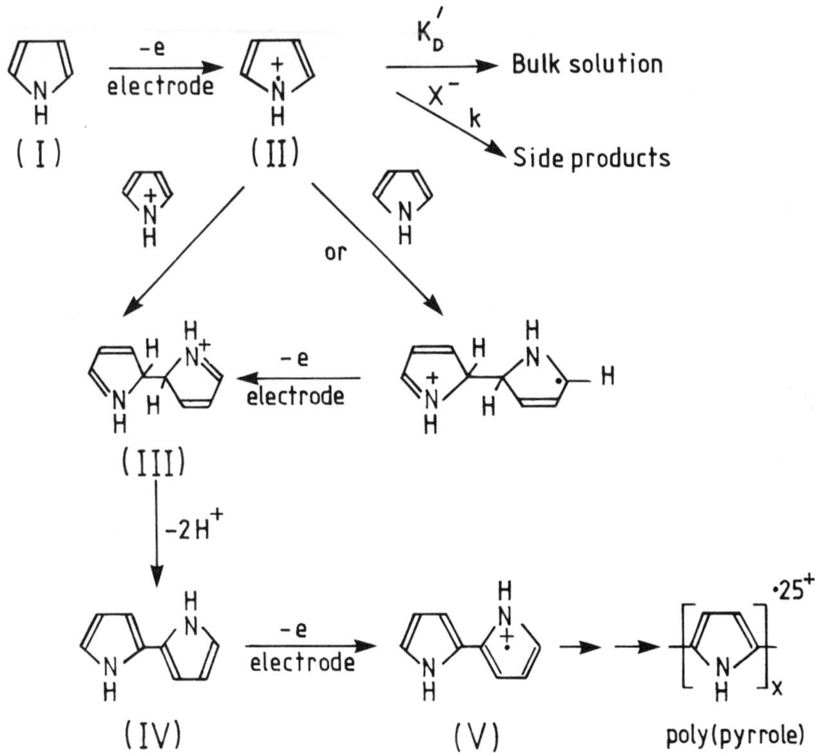

Figure 4.8 Reaction scheme for the polymerisation of pyrrole [4.22].

Figure 4.9 shows an electron micrograph of a thin poly(pyrrole) film that had been electrochemically grown over a microgap etched across a gold electrode. The poly(pyrrole) film was grown in an aqueous solution with toluene sulphonic acid as the counter-ion and forms microspheres as is clearly visible. A faint channel down the centre of the film is visible under which was milled across the underlying electrode using a focused ion beam technique (§4.3). The resulting structure is, in effect, a gas-sensitive resistor as the current that flows between the two electrodes has been found to be gas-sensitive [4.25]. The application of conducting polymers in gas sensors and odour sensors is discussed later on (§9.2.3).

4.5.5 Screen printing of thick films

The process of screen printing has been used for many years as a cheap and easy way to made hybrid circuits in electronics. The technique basically consists of the preparation of an ink paste using suitable organic solvents which is then squeezed through a fine gauze mask to leave a 25 to 100 μm film in the desired areas. The film is then dried by heat treatment to leave a conductive layer. The lateral resolution of the technique is only ~100 μm but the low printing cost often makes the technique commercially viable for low volume low-cost electronic circuits.

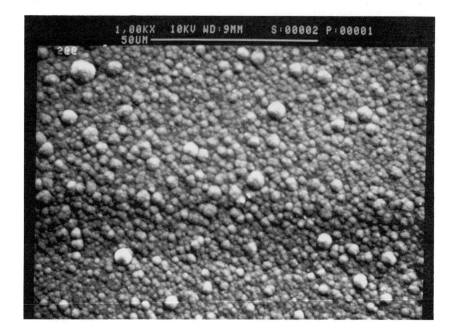

Figure 4.9   Micrograph of a poly(pyrrole) film lying across an ion beam milled electrode gap to form a gas sensor [4.25].

More recently, the screen printing process has been used to make microsensors. For example, platinum electrodes have been printed and used in electrochemical and bioelectrochemical sensors; thick metal phthalocyanine films also have been deposited on alumina to form the active material in a gas sensor [4.26]. Although this process often has a poor reproducibility compared to UV lithography, the low level of technology offers great commercial advantage in certain applications.

4.6 Comparison between Micromachining Techniques

As can be seen from this and the previous chapter, there is a considerable number of materials processing technologies that are used in the fabrication of microsensors. Specifically, we have considered the following principal methods:

- anisotropic and isotropic bulk silicon etching;
- surface silicon micromachining;
- reactive ion etching;
- physical ion etching;
- laser micromachining.

The nature of each of these micromachining techniques is now summarised in Table 4.7 so that a direct comparison is possible. The table also gives the type of materials that can be micromachined and their etch-rates. For example, the rate at which material is removed is clearly an important factor in any manufacturing process and can make one technology uneconomic. It can been seen that chemical wet etching can remove material quickly and over a large area which is of benefit. Each technology is also broadly classified in terms of its complexity - low, medium or high. This is a measure of the difficulty in setting up the process (and thus associated costs). Finally general comments are made upon the type of structures that can be micromachined and the performance of the technology.

Table 4.7 Summary of the principal micromachining technologies, their limitations and typical uses.

Technology	Material	Etch-rate (nm/min)	Complex	Uses, weaknesses and strengths
Anisotropic wet etching (75-120°C)	Si <100> Si <110> Si <111>	400-3,000 130-1,700 8-180	Low	Cusped pillars, angled mesa, nozzles, diaphragms, cantilevers, bridges. Imprecise control, use etch-stop.
Isotropic wet etching (25°C)	SiO_2 Si_3N_4 PSG	~70 ~1 ~1,000	Low	Spheres, domes, grooves. Imprecise control.
Reactive ion	Polymers	100-400	Medium	Cantilevers, bridges, holes, motors. Precise control.
SIBM[1] (500 eV)	Metals Resists	~50 ~30	High	Milling, sharpening of tips.
FIBM[1] (20 keV) (1 μm²)	Metals Resists	~1,000 ~1,000	High	Microholes, channels, VLSI circuit repair. Precise control, high resolution.
Laser (KrF) (2.0 J/cm²)	BSG Si_3N_4 Polyimide	~1500 ~1900 ~400	Medium	Circuit repair, masks, chips, PCBs Precise control, high etch-rate.
Electrostatic discharge	Metals	High	Low	Drilling holes and channels. Poor resolution, needs good conductors.

[1] It is also possible to machine other materials such as ceramics.

One of the major limitations in surface micromachining practice has been its restriction to relatively thin layers, i.e. a few microns. Although this is not usually applicable to microsensors, which often have thin active layers, it is often a limiting factor in the fabrication of microactuators. Microactuators often require large microstructures, such as microgears with a 100 μm diameter. The recent development of the LIGA process [4.27] has allowed the etching of structures to a depth of 100 μm or more. The LIGA process uses X-ray lithography with electroforming and moulding processes and enables the increased vertical depth to produce a device with a higher mechanical or electrical coupling. However, the cost of the synchrotron radiation source needed to generate the collimated beam of hard X-rays limits its implementation as a standard sensor manufacturing process. (Soft X-rays are used in high resolution lithography which are generated by cheaper and more compact storage rings). Nevertheless there is no doubt that considerable improvements will be made in micromachining technologies as microdevices become increasingly commercially successful.

Suggested further reading

The following books are recommended to readers who wish to know more about micromachining technology for microdevices, although there is no definitive book yet published in this field:

Gardner JW and Hingle HT (eds.): *From Instrumentation to Nanotechnology* (1991). Published by Gordon & Breach Science Publishers, Philadelphia. ISBN 2-88124-794-6. (336 pages. Several chapters give some introductory information on microengineering techniques)

Middelhoek S and Audet SA: *Silicon Sensors* (1989). Published by Academic Press, London. ISBN 0-12-495051-5. (376 pages. See chapter 7 for a discussion of silicon sensor technology)

References

4.1   von Engel A (1983) *Electric Plasmas: Their Nature and Uses*, Taylor and Francis Ltd, London.

4.2   Sugano T (ed.) (1985) *Applications of Plasma Processes to VLSI Technology*, John Wiley and Sons, New York.

4.3   Law VJ (1992) Plasma-etching of materials, in *Course notes on Micromachining of Materials*, University of Southampton, March 1992, p.43.

4.4   Law VJ, Tewordt M, Ingram SG and Jones GAC (1991) Alkane based plasma etching of GaAs. *J. Vac. Sci. Technol. B*, **9**, 1449-1455.

4.5   Law VJ and GAC Jones (1992) Chloromethane based reactive ion etching of GaAs and InP. *Semicond. Sci. Technol.*, **7**, 281-283.

4.6   Kaufman HR and Robinson RS (1989) Broad-beam ion source technology and applications. *Vacuum*, **39**, 1175-1180.

4.7   Davies ST and Whitehouse DJ (1987) Computer-controlled ion beam machining for precision surface finishing and figuring. *SPIE*, **803**, 37-42.

4.8   Somekh S (1976) Introduction to ion and plasma etching. *J. Vac. Sci. Technol.*, **13**, 1003-1007.

4.9   Davies ST (1989) Ion beam milling of diamond. *Industrial Diamond Review*, **49**, 201-204.

4.10   Melliar-Smith CM (1976) Ion etching for pattern delinearisation. *J. Vac. Sci. Technol.*, **13**, 1008-1022.

4.11   Wagner A, Levin JP, Mauer JL, Blauner PG, Kirch SJ and Longo P (1990) X-ray mask repair with focused ion beams. *J. Vac. Sci. Technol. B*, **8**, 1557-1564.

4.12   Melngailis J (1987) Focused ion beam technology and applications. *J. Vac. Sci. Technol. B*, **5**, 469-495.

4.13   Exitech Ltd, Long Hanborough, Oxford, UK. Technical data-sheet, 1992.

4.14   Chuang TJ (1986) Laser-induced molecular processes on surfaces. *Surf. Sci.*, **178**, 763-786.

4.15   Petersen KE (1982) Silicon as a mechanical material. *Proc. IEEE*, **70**, 420-456.

4.16   James AM and Lord MP (1992) *Macmillans Chemical and Physical Data*, Macmillans Press Ltd, London, UK, 565 pp.

4.17   Pitt CW (1983) Langmuir-Blodgett films - the ultrathin barrier. *Electronic and Power*, **29**, 226-229.

4.18   Imanpour M (1986) Development of processing conditions for production of Langmuir-Blodgett films for an electronic nose. *MSc dissertation, Warwick University, UK*.

4.19   Jones TA and Bott B (1986) Gas induced electrical conductivity changes in metal phthalocyanines. *Sensors and Actuators*, **9**, 27-37.

4.20   Roberts GG (1983) Transducer and other applications of Langmuir-Blodgett films. *Sensors and Actuators*, **4**, 131-136.

4.21   Rieger PH (1987) *Electrochemistry*, Prentice-Hall International, London, 508 pp.

4.22   Gardner JW and Bartlett PN (1991) Potential applications of electropolymerised thin organic films in nanotechnology. *Nanotechnology*, **2**, 19-32.

4.23   Salmon M, Diaz AF, Logan AJ, Kronubi M and Bargon J (1982) Chemical modification of conducting polymer films. *Mol. Cryst. Liq. Cryst.*, **83**, 1297-1308.

4.24   Skotheim TA (ed.) (1986) *Handbook of Conducting Polymers*, Marcel Dekker, New York.

4.25   Friel S, PhD thesis, University of Warwick, UK, in preparation.

4.26   White NM and Cranny AWJ (1987) Design and fabrication of thick film sensors. *Hybrid Circuits*, **12**, 32-35.

4.27   Ehrfeld W, Gotz F, Munchmeyer D, Schelb S and Schmidt D (1991) LIGA process: sensor construction techniques via X-ray lithography, in *Microsensors* (eds. R.S. Muller, R.T. Howe, S.D. Senturia, R.L. Smith and R.M. White), IEEE Press, USA.

Problems

4.1   Briefly describe which techniques could be used to etch the following materials: (a) a metal and (b) a polymer. If it was necessary to mass-produce a polymeric device using a micromachining technology, then which one would you use and why?

4.2   Calculate the etch-rate of aluminium using a beam of gallium atoms in FIBM. State clearly any assumptions that you make, and suggest a possible application for this process.

4.3   Describe a method by which the following active sensor materials can be deposited: (a) quartz, (b) iron and (c) poly(pyrrole). State, in turn, two types of microsensor that use each material together with their principle of operation.

4.4   An electronic engineer discovers that a metal interconnect in a VLSI device is incorrect. Could the engineer repair the routing error on-chip? If so describe a possible method that could be used.

5. Thermal Microsensors

Objectives

- ☐ To present the basic properties of thermal microsensors
- ☐ To understand the operation and principles of thermal microsensors
- ☐ To describe the relative importance of thermal microsensors

---

5.1 Basic Considerations and Definitions

5.1.1 Thermal measurands

Thermal sensors are used to measure various heat-related quantities, such as temperature, heat flux and heat capacity. Temperature is perhaps the most fundamental variable quantity and provides a measure of the thermal energy or heat in a body. The temperature of a body can also be related to the other thermal quantities.

The amount of heat Q (units of Joules) in a body is proportional to its absolute temperature T (in Kelvin),

$$Q = mcT \tag{5.1}$$

where the constant of proportionality mc is the heat capacity, m being the mass and c the heat capacity per unit mass or specific heat capacity (units of J/kg K) which is a fundamental property of the material. The specific heat capacity of a body is a measure of its ability to store thermal energy and is analogous to the capacitance of an electrical capacitor which can store electrostatic energy.

A difference in the temperature of two connected bodies or regions acts like a driving voltage so that heat will flow from the hotter region to the colder one. The rate dQ/dt at which heat flows through a region is determined by the temperature gradient dT/dx established and the thermal conductivity κ of the region, where

$$Q = \frac{dQ}{dt} = -\kappa A \frac{dT}{dx} \quad \text{or, more generally,} \quad -\kappa A \nabla T \qquad (5.2)$$

where A is the cross-sectional area of the region through which the heat flows, and ∇ is the 3-d grad vector operator.

The thermal resistance of a body R_T is defined as a measure of a body's ability to resist the flow of heat passing through it and is related to the thermal conductivity of the material and geometry of the body by,

$$R_T = \frac{L}{\kappa A} \qquad (5.3)$$

where L is the length of the body. Thus, temperature and heat flux are fundamental variables, equivalent to voltage V and current i in an electrical system, while heat capacity and thermal resistance describe the basic properties of a system and are equivalent to capacitance C and electrical resistance R. Table 5.1 shows the analogies between a thermal and an electrical system which can be used to help model and characterise the behaviour of a thermal system.[1]

Table 5.1 Analogy between a thermal and an electrical system.

System	Variables	Elements	Static law	Dynamic law
Electrical	i, V	R, C, L	$q = C/V$	$V = iR$
Thermal	Q, T	$R_T, c, -$	$Q = c/T$	$T = QR_T$

There is no known thermal equivalent to an electrical inductor L and, of course, the real behaviour of a thermal system is non-linear because the heat capacity and thermal resistance are generally temperature dependent. However, a linear model of a thermal system or a thermal sensor can often yield a good estimate of its characteristic behaviour. For example, a useful estimate of the real thermal time constant can be obtained from a simple model of the temporal response of a body as a first order system, i.e. an exponential function (§1.3).

5.1.2 *Application considerations*

Temperature is perhaps the most important process parameter and about 40% of all solid-state sensors sold in 1987 were thermal sensors (see Table 1.6). For instance, temperature is important in most chemical processes as the chemical reaction-rate is usually exponentially temperature-dependent according to the Arrhenius relationship (§9.1.2). So temperature is a fundamental parameter in many processes and it may need to be measured, compensated for, or even controlled in some manner. The industrial processing of many materials ranging

---

[1] The lumped system model is usually limited by an assumption of linearity.

from the traditional casting of steels and other metals to the micromachining of single-crystal silicon for microsensors (§3.3) requires the measurement of temperature. Not only is temperature often measured in plant, automobiles (e.g. the engine), household appliances (e.g. a refrigerator), medicine (e.g. body temperature) and environment (e.g. air temperature), but it is also exploited as a secondary sensing variable in non-thermal microsensors, such as a gas sensor (§9.6) or a mechanical wind flow sensor (§7.4.2).

Temperature sensors are also used to compensate for errors caused by the temperature variation of components or instruments, or form part of a control circuit (e.g. a thermostat). Thus thermal sensors form perhaps the largest and most important class of microsensors today.

5.1.3 Classification of thermal sensors

Thermal sensors are classified, by definition, as contacting, rather than non-contacting sensors, in which the sensing element physically touches the heat source. The thermal signal is then propagated from the heat source by conduction of heat into the sensing element which then usually either generates an electrical signal or modulates an electrical signal. Non-contacting temperature sensors are classified as radiation sensors (§1.2). Radiation temperature sensors detect the electromagnetic waves emitted by a body (e.g. a pyroelectric detector) and are discussed in Chapter 6.

Figure 5.1 shows a classification scheme for the family of thermal sensors. The majority of thermal electrical sensors are modulating rather than self-generating. The two exceptions are the thermocouple (or thermopile) which generates an e.m.f. across two junctions each being held at a separate temperature, and thermal noise sensors. The majority of the thermoconductive sensors, e.g. thermistors, thermodiodes and thermotransistors, can be classified as microsensors. However, conventional temperature sensors such as the mercury-filled thermometer are not microsensors and so will not be discussed in detail here. As can be seen from Table 5.2 contacting thermal microsensors can be used to cover a wide range of temperatures (i.e. from 1.5 to 2,700 K) whereas non-contacting pyroelectric radiation sensors can measure up to 5,000 K. The table also shows the temperature ranges covered by standard measurement techniques, and its typical resolution [5.1]. An advantage of using a liquid-filled thermometer or platinum resistance thermometer is that they have a nearly constant Temperature Coefficient of Resistance (TCR), i.e. an output that is directly proportional to temperature over a significant temperature range. However, the varying TCR of silicon and other materials is now only a slight disadvantage, because it is possible to make linearising analogue circuits or digital calibration routines at low cost.

The physical principles underlying each type of thermal sensor will now be considered in detail. This will help explain the limitations of each type of sensor and thus determine its appropriate application.

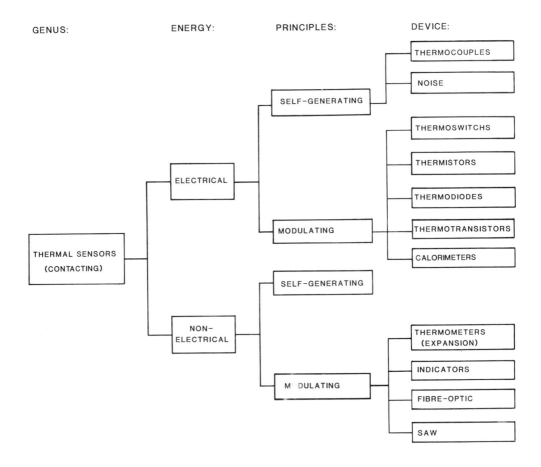

Figure 5.1 Classification scheme of thermal sensors.

Table 5.2 Range and resolution of various temperature sensors. Adapted from [5.1].

Sensor/technique	Range achieved (K)	Type	Resolution (K)
Microsensors:			
Germanium resistance thermometer	1.5 to 100	Laboratory	0.0001
Carbon resistance thermometer	1.5 to 100	Laboratory	0.001
Platinum resistance thermometer	15 to 1,000	Standard	0.00001
Thermistor	4 to 500	Laboratory	0.001
Silicon p-n junction	210 to 430	Laboratory	0.1
Thermocouple	20 to 2,700	General purpose	1.0
Radiation (§6.4)	270 to 5,000	Industrial	2
Traditional:			
Gas thermometer	1.5 to 1,400	Laboratory	0.002
Liquid-in-glass	130 to 950	General purpose	0.1
Bimetallic strip	130 to 700	Industrial	1 to 2

5.2 Thermocouples

5.2.1 Thermoelectric effects

When a circuit consists of two different metals (e.g. copper and iron) and the junctions are held at different temperatures, then an e.m.f. is generated and an electrical current will flow between them. Figure 5.2 shows the basic arrangement where a junction of two materials (e.g. copper and nickel) is held at a temperature T_A while a second (reference) junction is held at a different temperature T_B. A thermoelectric potential ΔV is generated across the junctions which can be measured by a voltmeter, usually with a high load resistor, R_L. The device is usually referred to as a *thermocouple* and the thermoelectric effect is known as the Seebeck effect. The open circuit voltage ΔV depends upon the difference in the temperature between the thermocouple T_A and the reference point T_B. It is related to the Seebeck coefficient or thermopower P_A which is material-specific and given by,

$$P_A = \frac{\Delta V}{\Delta T} \tag{5.4}$$

As a working thermocouple requires two metals, the Seebeck coefficient measured by a thermocouple P_s is

$$P_s = P_B - P_A \tag{5.5}$$

where P_B is the Seebeck coefficient of the reference metal B. Table 5.3 lists the thermoelectric e.m.f.'s ΔV_r generated by some typical metals and alloys relative to a platinum standard at a temperature of 0°C [5.2], where

$$\Delta V_r = (P_A - P_B)\Delta T \tag{5.6}$$

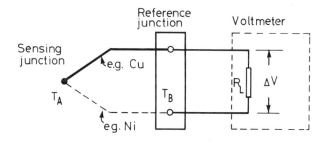

Figure 5.2 Basic circuit of a thermocouple temperature sensor.

Table 5.3 Thermoelectric e.m.f.'s and thermopowers of various metals and standard thermocouple alloys at 200°C relative to platinum at 0°C. From [5.2].

Element	ΔV_r (mV)	P_s (μV/K)	Element	ΔV_r (mV)	P_s (μV/K)
Antimony	+10.14	+50.7	Pt-10% Rh	+1.44	+7.20
Chromel	+5.96	+29.8	Aluminium	+1.06	+5.30
Iron	+3.54	+17.7	Tantalum	+0.93	+4.65
Molybdenum	+3.19	+16.0	Platinum	0.00	0.00
Tungsten	+2.62	+13.1	Calcium	-0.51	-2.55
Cadmium	+2.35	+11.8	Palladium	-1.23	-6.15
Gold	+1.84	+9.20	Alumel	-2.17	-10.85
Copper	+1.83	+9.15	Cobalt	-3.08	-15.40
Silver	+1.77	+8.85	Nickel	-3.10	-15.50
Rhodium	+1.61	+8.05	Constantan	-7.45	-37.25
Pt-13% Rh	+1.47	+7.35	Bismuth	-13.57	-67.85

From Table 5.3 it is possible to obtain the e.m.f. for any thermocouple by subtracting the two values listed. Ideally, a thermocouple should have one arm made of a high e.m.f. material, e.g. iron at +3.54 mV, and one with a low e.m.f. material, e.g. nickel at -3.10 mV. The resulting e.m.f. for a 200°C temperature gradient is thus 6.64 mV or a thermopower of 33.2 μV/K. The table also includes the mean thermopowers of the various materials over this temperature range; the thermopower is temperature-dependent.

The Seebeck effect in metals (and alloys) is quite small and its exact value is not easy to predict. However, the thermoelectric e.m.f. (or electrochemical potential) is normally associated with combined changes in the Fermi energy E_F and the diffusion potential. The Fermi level effect ΔV_F is given by

$$\Delta V_F = P_s \Delta T = \frac{\Delta E_F}{q} \tag{5.7}$$

where q is the electron charge. The Fermi level of a metal depends upon its temperature T and the density of states $N(E)$, and is given by

$$E_F(T) = E_F(0) - \pi^2 \frac{k^2}{6} T^2 \frac{d(\ln N(E))}{dE} \tag{5.8}$$

where $E_F(0)$ is the Fermi level of the metal at absolute zero and k is Boltzmann's constant. Thus, raising the temperature of a metal like nickel causes the Fermi level to fall. A lowering of the Fermi level at the hot junction leads to electrons flowing towards it and thus produces a positive Seebeck coefficient. However, the temperature gradient also affects the average velocities of both the electrons and the phonons within the metal. The differences in electron and phonon densities cause a diffusion potential to develop which opposes the e.m.f. calculated from the change in the Fermi level and thus reduces the overall observed effect.

Two other thermoelectric effects have also been discovered but are of less importance in microsensors. First, the Peltier effect occurs when a current is passed through a junction of dissimilar conductors and heat is either released or adsorbed depending upon current direction. Secondly, the Thomson (Kelvin) effect occurs when a current is passed along a single conductor and a temperature gradient is maintained, additional heat, besides Joule heat, is generated or adsorbed. However, as the Peltier and Thomson effects have not yet been exploited in sensors they will not be discussed further.

5.2.2 Thick wire, thin wire and foil thermocouples

The most widely used thermocouples have been standardised in both the United States and the United Kingdom. In the UK, the colour coding of thermocouples is specified in the British standard BS 1843 (1952) while the type designations (J, K, N, T, R) are specified in BS 4937 by their composition, temperature range, accuracy, etc. Different colour coding by similar type designations have been specified by the American standard ANSI MC 96 (1975).

Table 5.4 gives some of the specifications of standard UK thermocouples according to BS 4937. For each thermocouple type, the materials used, minimum and maximum temperatures and accuracy are specified.

Three common alloys are used in thermocouples, namely, Chromel (90% Ni, 10% Cr), Constantan (55% Cu, 45% Ni) and Alumel (95% Ni, 2% Al, 2% Mn, 1% Si). Nicrosil (71-86% Ni, 14% Cr, 0-15% Fe) and Nisil (mainly 95% Ni, 4.5% Si) alloys are used in a Type N thermocouple which has a similar output to the Type K thermocouple but possesses a better stability.

Figure 5.3 shows the typical thermoelectric e.m.f.'s generated by ANSI Type E, J, K, T, R, S and B thermocouples over a temperature range of 0 to 1,000°C. The materials used in each type are Chromel/Constantan (E), Iron/Constantan (J), Chromel/Alumel (K), Copper/Constantan (T), Pt/Pt-13%Rh (R), Pt/Pt-10%Rh (S) and Pt-6%Rh/Pt-30%Rh (B). The reference junction is held at 0°C while the other

Table 5.4 Specification of UK standard metal/alloy thermocouples.

Parameter	Type J	Type K	Type N	Type T	Type R
Minimum temp. (°C)	-40	-200	-230	-250	-50
Maximum temp. (°C)	+850	+1,100	+1,230	+400	+1,350
Accuracy, Class 2 (°C)	±2.5	±2.5	±2.5	±1	±2
Temp. range, Class 2 (°C)	-40 to +750	-40 to +750	-40 to +750	-40 to +750	-15 to +1,760
Composition +ve arm	Iron	Chromel	Nicrosil	Copper	Platinum
Composition -ve arm	41% Ni/ 55% Al	Alumel	Nisil	Constantan	87% Pt/ 13% Rh

Table 5.5 Specification of US standard metal thermocouples (MC 96, 1975).

Parameter	Type B	Type E	Type J	Type K	Types R, S, T
Temperature range (°C)	800 to 1,700	0 to 900 -200 to 0	0 to 750	0 to 1,200 -200 to 0	0 to 1,450 0 to 350 -22 to 0
Accuracy (°C)	-	±1.7 ±1.7	±2.2	±2.2 ±2.2	±1.5 ±1.5 ±1.5
Accuracy (%)	±0.5	±0.5 ±1	±0.75	±0.75 ±2	±2.5 ±0.75 ±1.5

junction temperature is shown. As can be seen the output is reasonably linear with temperature. The specification of these US thermocouples is shown in Table 5.5.

Problems due to ferromagnetism may arise when using the Type K thermocouples in which the negative arm is ferromagnetic at room temperature and its Curie point lies within the operating range. Thus at high temperatures (200 to 600°C) some hysteresis occurs, and at very high temperatures the arms can oxidise causing a change in the e.m.f. output. The Type N thermocouple overcomes most of these problems and is recommended when available.

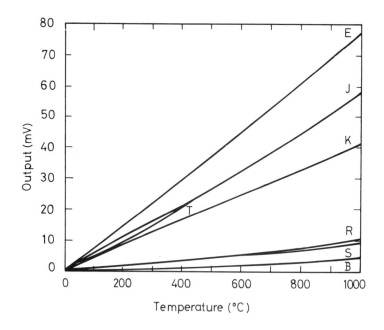

Figure 5.3 Thermoelectric e.m.f. output of standard thermocouples [5.2]. (Reproduced by permission of Prentice Hall Inc., © 1989).

Thick wire thermocouples are very robust and generally consist of a single wire of diameter 200 µm. Consequently, these are not microsensors as defined in this book. However, fine wire thermocouples are available and can respond rapidly ($\tau < 0.1$ s) to temperature changes due to their lower thermal mass. Ultrafast response ($\tau \sim 5$ ms) can also be obtained from thin foil thermocouples over 0 to 800°C. A typical foil thickness is 50 µm and is supported by a polyamide film. Thin foil thermocouples have the advantage that they can be easily attached to surfaces and occupy little space.

When a higher voltage output is required, a group of thermocouples can be wired in series. The resulting structure which was first used to detect radiant energy is called a *thermopile*. Figure 5.4 shows a thermopile arrangement of four identical thermocouples connected in series. The output is four times that of a single thermocouple. Recent advances in microelectronic microtechnology have resulted in the development of thin film thermopiles (Copper/Constantan and Nickel/Chrome) that may replace the thin foil thermocouples. Figure 5.5 shows the output of a Type T microthermopile with 90 thermojunctions for differential temperatures of up to 12 K. The figure also shows its power output when used in a microcalorimeter (see §9.6.3 on chemical microsensors). The observed Seebeck coefficient or thermoelectric power P_s is 25.3 µV/K for each junction or 2.28 mV/K for the device. The non-linearity over this temperature range is very good at 0.1% [5.3].

The thin-film deposition of some thermocouple materials can be difficult, especially when a particular composition is required with a good stability. Consequently, the majority of thermoelectric microsensors employ a semiconducting material such as silicon or germanium.

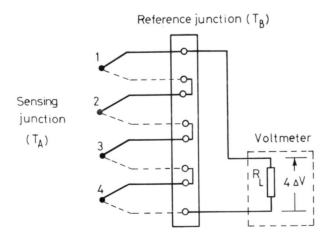

Figure 5.4 A basic thermopile circuit consisting of four identical thermocouples.

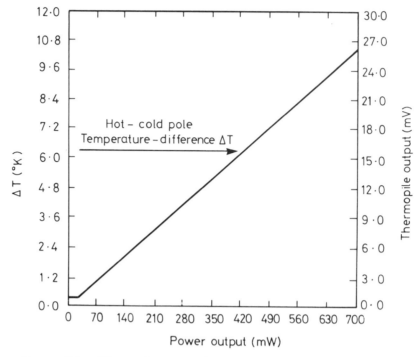

Figure 5.5   Thermoelectric and power output from a Type T microthermopile used in a microcalorimeter [5.3].

5.2.3 Semiconductor thermocouples

Semiconducting materials often show a thermoelectric effect which is larger than that observed in metals. As previously stated, the Seebeck effect is a bulk transport property that can be expressed in terms of the temperature gradient and change in Fermi level (from equation (5.7))

$$\nabla E_F / q = P_s \nabla T \qquad (5.9)$$

where E_F/q is the electrochemical potential ϕ and P_s the Seebeck coefficient (i.e. thermopower). For a non-degenerate semiconductor, the Seebeck coefficient can be estimated using Maxwell-Boltzmann statistics. There are three main effects that result in three separate terms, each of which modifies the electrochemical potential [5.4]. In an n-type semiconductor, the Seebeck coefficient is given by equation (5.10). The first term accounts for the material becoming more intrinsic as the gap between the Fermi level and the conduction band changes with temperature, where n is the doping level, N_c is the density of states at the conduction band, k is Boltzmann's constant and q is the electron charge. The second term arises from the charge carriers having an increasing average velocity as the temperature increases. In addition, the scattering of the charge carriers is usually energy-dependent and s_n is the exponent relating the mean free time between collisions and the charge carrier energy. s_n takes a value typically between -1 and +2 depending on whether the hot carriers can move more freely or are trapped by increased scattering. Lastly, the temperature differential causes a net flow of phonons from the hot region to the cold region. The net phonon movement from the hot to cold region causes a drag on the charge carriers represented by ϕ_n. The size of the phonon drag varies with doping level and temperature. Typically, ϕ_n has a value of zero in highly-doped silicon and 5 in lowly-doped silicon at 300 K.

A similar equation can be obtained for a p-type non-degenerate semiconductor to produce both equations as,

$$n\text{-type: } P_s = -\frac{k}{q}\left\{ \left(\ln\left(\frac{N_c}{n}\right) + \frac{3}{2} \right) + (1 + s_n) + \phi_n \right\}$$

$$(5.10)$$

$$p\text{-type: } P_s = +\frac{k}{q}\left\{ \left(\ln\left(\frac{N_v}{p}\right) + \frac{5}{2} \right) + (1 + s_p) + \phi_p \right\}$$

where N_v is the density of states at the top of the valence band and p is the acceptor concentration.

In practical situations, the Seebeck coefficient can be usefully related to the electrical resistivity ρ at various temperatures of interest in the following approximation:

$$P_s \approx \frac{mk}{q}\ln(\rho/\rho_0)$$

$$(5.11)$$

where m is a dimensionless constant with a value of about 2.6 [5.4] and ρ_0 is a resistivity constant of 5×10^{-6} Ω m. Figure 5.6 shows a plot of some experimental values of the Seebeck coefficient of single-crystal silicon at 300 K versus its electrical resistivity at various doping levels [5.5]. The dashed line shows a theoretical fit of equation (5.11) to the experimental data which is best at higher doping levels.

It is immediately obvious from Figure 5.6 that the Seebeck coefficient of silicon (~1,000 μV/K) greatly exceeds those values previously tabulated for various metals (~10 μV/K). This general observation is illustrated well when considering the "figure of merit" Z defined by

$$Z = P_s^2\sigma/\kappa$$

$$(5.12)$$

where σ is the electrical and κ is the thermal conductivity of the material. The figure of merit depends not only upon the size of the Seebeck effect but also the ratio of electrical to thermal conductivity. A high conductivity ratio is advantageous because it permits a high temperature gradient to be maintained at a low power consumption. Figure 5.7 illustrates the potential advantage of using a semiconducting rather than a metallic (or insulating) material in a thermoelectric

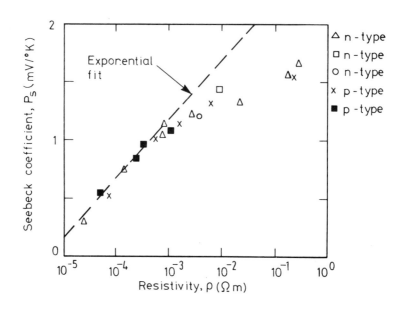

Figure 5.6 Observed Seebeck coefficients of single-crystal silicon for several resistivities [5.5].

sensor [5.6].   Although a somewhat crude model, it indicates that a peak performance occurs at a carrier density of between 10^{18} and 10^{20} cm^{-3}.

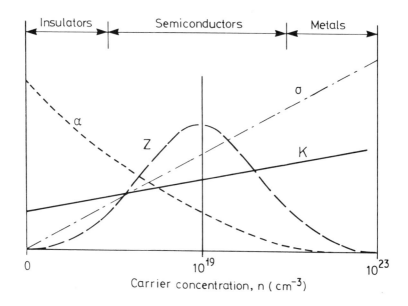

Figure 5.7 Figure of merit parameter Z for insulators, semiconductors and metals vs. carrier concentration *n* [5.6].

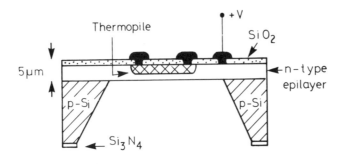

Figure 5.8 Micromachined silicon thermopile. Redrawn from [5.5].

Early work on microthermopiles showed that it was possible to fabricate devices using narrow ion-implanted p-type strips onto which are deposited aluminium connectors [5.7]. The major problem has been the high thermal conductivity of silicon producing a low device sensitivity. However, as has been recently reported, the bulk micromachining of a silicon substrate allows a thin membrane of perhaps only 5 to 10 µm to remain as illustrated in Figure 5.8 [5.5]. The thin membrane greatly increases the thermal resistance of the device and hence its sensitivity. The high thermoelectric power (~5 mV/K) of this silicon device demonstrates its the potential use in temperature sensing. However, integrated thermopiles have other more promising applications such as flow sensing (§7.4.2).

5.3 Noise Thermometry

Thermocouples form the largest group of electrical self-generating thermal sensors but, perhaps surprisingly, the phenomenon of electrical noise can be used as a self-generating signal. Random noise in a passive element, such as a resistor, is often referred to as thermal noise because its magnitude depends on temperature.

The thermal noise in a resistor of size R is equivalent to an r.m.s. voltage V_n in series with the resistor and is defined by,

$$V_n = \sqrt{(4kR\Delta f)T}$$
(5.13)

where Δf is the frequency band-width of the circuit, k is Boltzmann's constant and T the absolute temperature. Typically, the thermal noise of a 10 kΩ resistor at a temperature of 290 K and a band-width of 10 kHz is about 1.3 µV. This small voltage limits its use as a thermal sensor.

Thermal noise can also be considered as a current generator where the current i_n is related to the conductance G by,

$$i_n = \sqrt{(4kG\Delta f)T} \qquad\qquad (5.14)$$

The resistor should be connected to a high-gain amplifier which does not contribute significant noise of its own. Thermal noise thermometers need careful calibration and operation, yet some success has been achieved through the use of superconducting Josephson junctions as preamplifiers to remove the amplifier noise.

Like resistors, both diodes and transistors can also be regarded as thermal noise generators. In these cases there are additional noise mechanisms such as Shot noise, Partition noise and Flicker noise. Shot and Partition noise arise from random fluctuations in the movement of minority and majority carriers. Figure 5.9 shows the equivalent circuit for the noise generated by a typical transistor. There is Shot noise current i_{en} in the emitter, Shot and Partition noise current in the collector i_{cn}, and thermal noise voltage due to the base resistance R_b where,

$$i_{en} = \sqrt{2e.i_e.\Delta f}$$

$$i_{cn} = \sqrt{2e(i_{co} + i_c(1-\alpha_0)\Delta f)} \qquad\qquad (5.15)$$

$$V_{bn} = \sqrt{4kT.R_b.\Delta f}$$

and α_0 is a temperature correction factor. Flicker noise is believed to be due to carrier generation and recombination at the base-emitter surface and is only important at low frequencies (i.e. ≤ 1 kHz).

The complex nature of the temperature dependence of noise in diodes and transistors, coupled with the inherently small signals generated, makes this type of sensor rather less practical than other types of temperature microsensors.

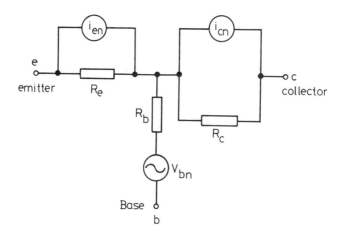

Figure 5.9 Equivalent circuit of noise in a transistor.

5.4 Thermoresistors

5.4.1 Metal thermoresistors

Although the electrical resistivity of both metals and semiconductors varies with temperature, the nature of their behaviour is different. The behaviour of a metal is illustrated in Figure 5.10 where the resisitivity of platinum is plotted over the temperature range of 0 to 500 K. At very low temperatures (<20K), the resistivity is typically constant, independent of temperature and governed by electron-impurity scattering (except in superconductors). At low temperatures (*ca.* 20 K to 50 K), a power law is observed which relates to an electron-electron scattering process. At temperatures above about 50 K, the relationship is nearly linear and is determined by electron-phonon scattering. It is over this region, i.e. -200°C to +1,000°C, that metal thermoresistors (or thermistors) can be employed.

The temperature coefficient of resistivity α_r is defined as,

$$\alpha_r = \frac{1}{\rho_0}\frac{d\rho}{dT} \qquad (5.16)$$

For metals in the approximately linear region, the resistivity ρ can be well described by a second order polynomial,

$$\rho \approx \rho_0(1 + \alpha T + \beta T^2) \qquad (5.17)$$

where ρ_0 is the resistivity at a standard temperature of 0°C and α and β are material constants. From equation (5.16), the temperature coefficient of resistivity is given by,

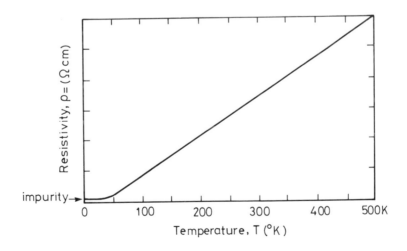

Figure 5.10 Electrical resistivity of platinum vs. temperature.

$$\alpha_r \approx \alpha + 2\beta T \tag{5.18}$$

Thus, the material constant α is the linear temperature coefficient of resistivity and for metals is positive and takes a typical value of ~5×10^{-3} /K (see Table 5.6).

Platinum is a popular choice of metal because its resistance is particularly linear with temperature (i.e. β is small with a value of -5.9×10^{-7} /K^2). Table 5.6 shows its resistivity relative to other common metals and alloys. Although platinum does not have the highest resistivity which would be advantageous in microdevices, it does have a reasonably high TCR (39.2×10^{-4} /K). More importantly, high purity wire is readily available and it has a high chemical stability. Wire-wound platinum resistance thermometers are widely used and industrial versions conform to the British Standard BS 1904 (1964). These thermometers cover a wide temperature range of 15 K to 1,000 K with a typical accuracy of 0.1°C and a nominal resistance of 100 Ω. A modified form of equation (5.17) called the Callendar-van Dusen equation is is used to calibrate platinum-wire elements. This equation extends the polynomial to fourth order (terms in T^4). Its stability enables platinum to be used as a temperature reference standard. Nickel resistance thermometers provide a cheaper alternative to platinum but only cover the range from 70 K to 600 K.

Table 5.6 Resistance characteristics of common metals and alloys.

Material	Resistivity, ρ (10^8 Ω.m) at 20°C	TCR, α_r (10^{-4}/K)
Nichrome (60% Ni, 16% Cr, 24% Fe)	109.0	2
Constantan (55% Cu, 45% Ni)	49.0	±0.2
Manganin (86% Cu, 12% Mn, 2% Ni)	43.0	−0.2
Palladium	10.8	37.7
Platinum	10.6	39.2
Iron	9.71	65.1
Indium	9.00	47.0
Nickel	6.84	68.1
Tungsten	5.50	46.0
Rhodium	4.70	45.7
Aluminium	2.69	42.0
Gold	2.30	39.0
Copper	1.67	43.0
Silver	1.63	41.0

Thin film (~1 µm thick) platinum themoresistors are a low-cost microsensor version of the platinum wire thermometer and possess lower thermal response times. Commercial sensors achieve an accuracy of less than ±0.2°C over 70°C to 350°C with a long term stability of better than 0.1°C per year.

The low nominal resistance (~100 Ω) and moderate TCR of metal thermoresistors necessitate the use of a resistance bridge network to transduce the signal. A 3-wire compensation bridge is often used to measure their resistance. Figure 5.11 shows a schematic arrangement of a 3-wire compensation bridge. There are three leads with resistances R_{L1}, R_{L2}, R_{L3} which connect the thermoresistor of resistance R_T to the bridge circuit. At balance, no current flows down the lead L_2 and, provided that the lead resistances can be made equal, there will be equal voltage drops across the other two leads L_1 and L_3. The bridge has maximum sensitivity when all four arms (i.e. R_1, R_2, R_3 and R_T) have an equal resistance. The bridge output V_{out} is about 1 mV/°C for a standard 100 Ω platinum thermoresistor and a 10 V DC voltage. The performance of electrical bridges has been discussed in Chapter 2 along with other signal processing schemes.

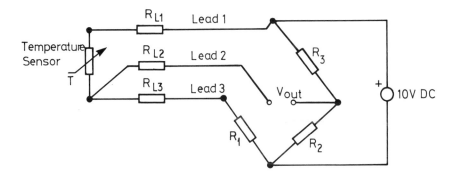

Figure 5.11 A three-wire compensation bridge circuit for metal thermoresistors.

5.4.2 Thermistors

It is possible to replace the metal in a thermoresistor by a temperature sensitive semiconducting resistor formed from the oxides of various metals or from silicon. These semiconducting thermoresistors are less stable and accurate than platinum thermoresistors but offer the advantages of a lower manufacturing cost and possibility of integrated interface circuitry. The word *thermistor* is generally used to classify semiconducting thermoresistors that are made from ceramic materials. The semiconducting materials (e.g. sulphides, selenides, oxides of Ni, Mn, Cu etc.) are formed into small discs, beads or rods. The resistivity of a typical thermistor is much higher than that of a metal thermoresistor and its TCR is negative and highly non-linear as illustrated in Figure 5.12. The plot shows the material resistances relative to its ice-point resistance in order to normalise the values for platinum and nickel.

The resistivity ρ of a thermistor is normally expressed at zero power by,

$$\rho = \rho_{ref} \exp\left[\beta(1/T - 1/T_{ref})\right] \tag{5.19}$$

where ρ_{ref} is the resistivity at a reference temperature T_{ref} (often 25°C rather than 0 °C), and β is a material constant. Thus, from equation (5.16), the TCR of a thermistor is

$$\alpha_r = -\frac{\beta}{T^2} \tag{5.20}$$

Typical values of the exponential constant β are 3,000 to 4,500 K and the corresponding zero power resistance at 25°C can be between 500 Ω and 10 MΩ.

The large change in resistance with temperature permits the use of thermistors with basic operational amplifier circuits rather than a precision bridge network. For example, a potential divider with a standard resistor R and a voltage amplifier (e.g. with a 741 op-amp) to buffer the signal may be adequate (see Figure 5.13), where the output voltage V_{out} is given by,

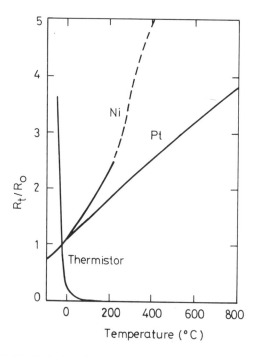

Figure 5.12 Typical plot of resistance vs. temperature of a thermistor.

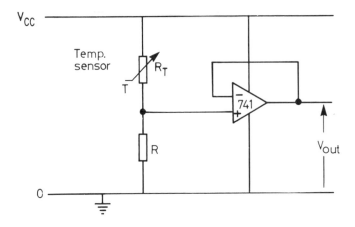

Figure 5.13 Basic circuit for a thermistor.

$$V_{out} = \frac{R}{R + R_T} . V_{cc} \tag{5.21}$$

The non-linear output with temperature is then usually calibrated using a look-up table in software.

Care is needed when operating thermistors because the TCR is strongly dependent upon the power consumption. There are two principal types of current-voltage characteristics as shown in Figure 5.14. An NTC thermistor (e.g. type BM, Bowthorpe) has a negative temperature coefficient of resistance so when a significant current flows, self-heating occurs, which leads in turn to a higher current at the same applied voltage.

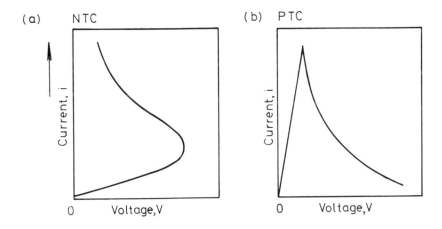

Figure 5.14 Current-voltage characteristics of NTC and PTC thermistors.

Some metal oxide materials possess a positive rather than a negative temperature coefficient of resistance and these PTC thermistors have a very different current-voltage characteristic. Now the high current flow causes the resistance to increase further, thus in turn reducing the current flowing through the device. The potential applications of these two types of thermistors are discussed later (§5.9.1). Thermal time constants (τ) of bead or disk thermistors, which are usually encapsulated in inert and stable glass or ceramic compounds, are typically 1 to 10 s in liquid.

5.4.3 *Semiconducting thermoresistors*

The use of intrinsic semiconductors as thermoresistors is attractive because the resistivity is theoretically governed by the carrier generation mechanism rather than the mobility term. Neglecting the weaker mobility temperature-dependent term, the resistivity ρ of a semiconductor may be expressed by,

$$\rho = A(T^n)\exp\!\left(E_g(0)/2kT\right) \;\propto\; \exp\!\left(E_g(0)/2kT\right) \tag{5.22}$$

where $E_g(0)$ is the energy band gap at absolute zero and $A(T^n)$ is a weakly temperature-dependent term with the exponent n being material dependent (~1 for silicon). Thus, the TCR of an intrinsic semiconductor is given by

$$\alpha_r \sim -E_g(0)/2kT^2 \tag{5.23}$$

Unfortunately, crystalline silicon does not exhibit an intrinsic behaviour because the lowest impurity levels obtainable (~10^{12} cm^{-3}) still exceed the intrinsic carrier concentration (~10^{10} cm^{-3} at room temperature). Germanium is used as a cryogenic temperature sensor, i.e. in the range of 1 to 35 K, but the ion impurity level required has effectively limited the application of intrinsic semiconducting thermoresistors to low temperatures. An alternative solution to this problem is to use an amorphous rather than crystalline intrinsic semiconducting material. A recent report has shown that amorphous germanium has a quasi-intrinsic behaviour which yields a high TCR [5.8]. A microdevice has been manufactured which has a long-term stability of 0.25°C over a range of 10°C to 60°C. The microdevice had a thermal time constant of only about 3 ms in liquid where the active layer was passivated by a 3 μm silicon nitride layer.

Perhaps a more practical approach is to use extrinsic rather than intrinsic silicon as a thermoresistor. The temperature region of most interest is between about 100 K and 500 K. At lower temperatures (<100 K), the resistivity falls with increasing temperature as carrier generation dominates. At high temperatures (>500 K), the resistivity also falls with increasing temperature as electrons are excited up to the conduction band. However, in the temperature range of 100 K to 500 K, the number of electrons in the conduction is determined by the doping level and the temperature-dependence is governed entirely by the carrier mobility. Figure 5.15 shows the resistivity of silicon versus temperature in this region at various doping levels [5.9]. At high doping levels (~10^{18} cm^{-3}), the resistivity is almost independent of temperature as impurity scattering exceeds phonon scattering. At moderate doping levels, the n-type silicon behaviour is quasi-metallic with a positive TCR and a resistivity at low temperatures that can be expressed, like metals, as a second order polynomial,

$$\rho \approx \rho_0 (1 + \alpha T + \beta T^2) \qquad (5.24)$$

where the constant ρ_0 is 1 to 10 Ω cm, and α and β are material constants. This behaviour has been exploited in silicon thermoresistors using an n^+/n contact to form a "spreading resistance" contact with a diameter of about 20 μm. Figure 5.16 shows the basic layout of such a device made using conventional microelectronic technology. The working range of this device can be extended through the application of a bias current. The bias current raises the temperature at which the semiconductor switches from extrinsic to intrinsic behaviour. Thus, at a current of 10 mA the maximum temperature increases from 500 K to about 700 K. The spreading resistance R_s is given by,

$$R_s = \frac{\rho}{\pi d} \qquad (5.25)$$

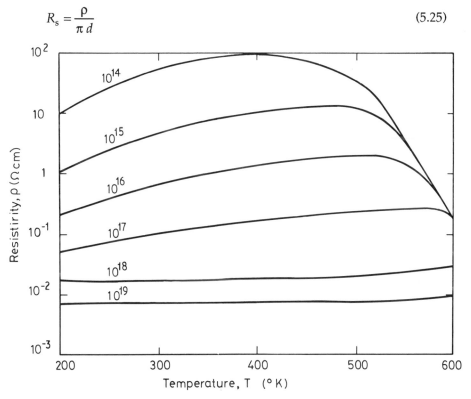

Figure 5.15 Temperature dependence of silicon at various n-type doping levels [5.9]. (Reproduced by permission of Pergammon Press Ltd.)

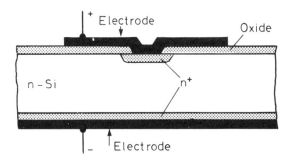

Figure 5.16 General layout of a silicon spreading resistance thermoresistor.

where d is the diameter of the circular contact.

Silicon temperature sensors such as these are manufactured by Siemens and typically have a resistance of 1 kΩ at 25°C with temperature coefficients α of 7.68 10^{-3} /K and β of 1.88 $\times$ 10^{-5} /K^2. A typical working range is from -50°C to +150°C with a thermal time constant τ of ~10 s (in air) or ~1 s (in liquid). The base current is up to 5 mA. Silicon thermoresistive sensors offer not only direct temperature measurement but also a much lower cost than thermistors.

5.5 Thermodiodes

5.5.1 Basic principles

The application of diodes and transistors as modulating thermal sensors was first proposed in 1962 by McNamara [5.10]. Potential advantages of thermodiodes over other types of thermal sensors include their compatibility with IC technology and low manufacturing cost.

Figure 5.17 shows the ideal current-voltage characteristic of a silicon *p-n* diode. At voltages less than the forward conduction voltage ($V<V_{f0}$), practically no current flows through the diode with a saturation current reached at higher reverse biases. At voltages above the forward voltage ($V>V_{f0}$), the diode is capable of passing a substantial amount of current.

The energy level diagram of a *p-n* junction under forward and reverse bias is shown in Figure 5.18 together with the depletion region (shaded area). At a zero bias voltage, the carriers must cross a potential barrier V_0 when moving from *n*-type region into the *p*-type region or vice versa. The height of this barrier is either reduced by the application of a forward bias voltage V_f or increased by the application of a reverse bias voltage V_r; in turn these effects modulate the current flowing through the junction. The Fermi level is modified by the applied voltage V as shown but always lies near the valence band in a *p*-type or near the conduction band in an *n*-type semiconducting material.

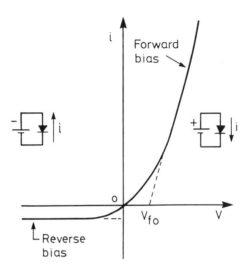

Figure 5.17 Current-voltage characteristic of an ideal silicon *p-n* diode.

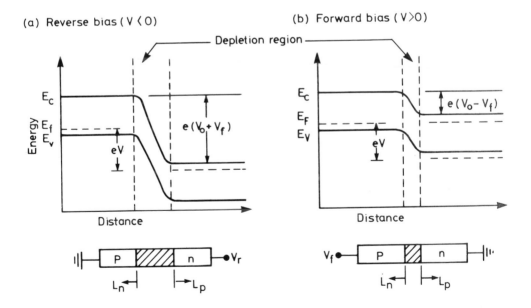

Figure 5.18 Energy level diagrams of a silicon *p-n* diode under (a) a reverse and (b) a forward bias voltage. The depletion regions are illustrated as shaded areas.

In a classical junction theory, the electron and hole current densities, J_n and J_p, can be related to the process of simple diffusion by,

$$J_n = D_e\, e(dn/dx) = D_e\, e\left[\left(n'_p - n_p\right)/L_p\right]$$

$$J_p = D_h\, e(dp/dx) = D_h\, e\left[\left(p'_n - p_n\right)/L_n\right]$$

(5.26)

where D_e is the diffusivity of the electrons, n'_p is the electron concentration at the depletion layer edge in the p-type material, and n'_p is the equilibrium electron concentration at a distance L_p from the edge. Similar parameters can be defined for the n-type region. Using Einstein's relationship, the electron and hole concentrations can be related to the external voltage V by,

$$\frac{n'_p}{n_p} = \exp\left(\frac{eV}{kT}\right) \text{ and } \frac{p'_n}{p_n} = \exp\left(\frac{eV}{kT}\right)$$

(5.27)

Substituting into equation (5.26) for the electron and hole concentrations gives the ideal diode equation (as plotted in Figure 5.17) as,

$$i = i_s\left[\exp\left(\frac{eV}{kT}\right) - 1\right]$$

(5.28)

where the saturation current is a constant i_s which is related to the junction area A and previously defined junction parameters by,

$$i_s = A\left(\frac{D_e e n_p}{L_p} + \frac{D_h e p_n}{L_n}\right)$$

(5.29)

In real diodes, the exponential term in the ideal diode equation (5.28) is modified by a non-ideality multiplicative factor η_d (typically having a value of 1.5) because thermally-generated carriers can diffuse into the depletion region and recombine there. In addition, effects such as a voltage drop across the neutral region or the diode package can reduce the actual voltage across the junction from that applied.

The ideal temperature dependence of the diode voltage can be derived by rearranging equation (5.28) to give,

$$V = (kT/e)\ln(i/i_s + 1)$$

(5.30)

This relationship is excellent for a temperature sensor as the diode voltage is directly proportional to the absolute temperature.[2] Operating the diode via a

---

[2] This type of thermal sensor is known as a PTAT device.

constant current source gives a temperature coefficient (dV/dT) that is theoretically independent of temperature and only depends upon the current terms, see equation (5.31).   Under a forward bias $i \ll i_s$, variations in the saturation current become negligible.

$$\frac{dV}{dT} = \frac{k}{e} \ln\left(\frac{i}{i_s} + 1\right) \quad \propto \quad \frac{k}{e} \ln(i) \qquad (5.31)$$

A forward junction voltage V_{f0} can be estimated from the current-voltage characteristic of a diode under forward bias (see Figure 5.17), although there is no true junction voltage as the derivative of the function is continuous. The forward junction voltage of silicon is 0.7 V at 25°C (cf. 0.25 V for germanium) and is approximately linear with temperature with a slope of -2 mV/°C.

Figure 5.19 shows a plot of the typical forward junction voltage of a silicon diode against temperature with a forward current of 10 μA. An approximately linear region is apparent with a sharp increase in forward voltage at very low temperatures.   Thus the forward junction voltage can be used to measure temperatures between 50 and 300 K.

Silicon thermodiodes first become non-linear in their forward voltage-temperature characteristic and then fail at low temperatures, because the doping atoms are no longer fully ionised. Conversely, at temperatures above about 200°C the mobility of the doping atoms is enhanced by temperature causing permanent damage to the junction.   Nevertheless over this limited temperature range thermodiodes offer a low cost way of measuring temperature to a moderate accuracy.

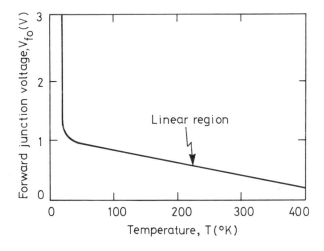

Figure 5.19 Variation of the forward junction voltage of a silicon diode with temperature (i=10 μA).

5.5.2 Integrated circuit thermodiodes

Figure 5.20(a) shows a basic op-amp circuit that could be used to measure the forward bias voltage V_f dropped across a thermodiode. The resistors R_1 and R_2 limit the current i which flows through the junction. The forward voltage across the diode is then measured via a differential voltage operational amplifier to give an output V_{out} linearly temperature-dependent. The commercially-available National Semiconductor device LM3911 is a low-cost temperature microsensor which contains an integral amplifier and operates from a single voltage input and has an output of 10 mV/K, see Figure 5.20(b). In practice, the saturation or leakage current i_s is sensitive to the manufacturing process and is itself affected by the diffusion and recombination rates and thus temperature. The saturation current for a silicon diode is only 25 nA at 25°C but it reaches a level of about 7 mA at 150°C. This high current demand limits both the range and linearity of thermodiodes.

The performance of an IC thermal microsensor can be improved upon through the use of a thermotransistor rather than a thermodiode, with the circuits containing laser-trimmed resistors to enhance its accuracy. Moreover, the negative temperature coefficient of the forward junction voltage from thermodiodes is undesirable in many control circuits because an open-circuit failure will lead to the application of full power.

(a)   (b)

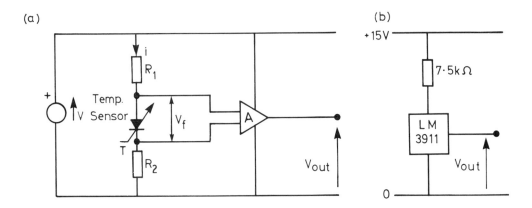

Figure 5.20   Basic circuits for (a) a silicon thermodiode and (b) a National Semiconductor commercial temperature IC.

5.6 Thermotransistors

5.6.1 Basic principles

Thermotransistors offer potential benefits over other types of temperature sensors because they are more easily integrated into microelectronic circuits. The emitter current i_e in a transistor is determined mainly by a diffusion process, rather than the recombination or leakage processes which determine the base current. The voltage between the base and emitter of the transistor, V_{be}, is temperature-dependent with a relationship similar to the diode expression stated earlier. The only difference is that the constants (e.g. i_s) have a different dependence upon the device geometry.

A more sophisticated method to measure temperature with one transistor is to apply first a high collector current i_{c1} and then secondly a low collector current i_{c2}. The difference in base-emitter voltages ΔV_{be} now only depends upon the collector currents rather than the geometrical or material factors,

$$\Delta V_{be} = (V_{be1} - V_{be2}) = \frac{kT}{e}\ln\left(\frac{i_{c1}}{i_{c2}}\right) \text{ with } i_{c1} \gg i_{c2} \tag{5.32}$$

This device is a PTAT like the thermodiode described earlier as the output is directly proportional to the absolute temperature.

5.6.2 Integrated devices

Figure 5.21 shows a circuit containing several thermotransistors which is more robust than a single thermoresistor circuit. Transistors T_1 and T_2 have the same base-emitter voltage V_{be1} and so the total current i splits equally down each side. In the second stage T_3 consists of eight transistors all identical to T_4. The collector current passing through transistor T_4 is thus eight times that passing through T_3. The voltage dropped across the resistor R is the difference between the base-emitter voltages, i.e.

$$V_R = (V_{be4} - V_{be3}) \tag{5.33}$$

and substituting in for the base-emitter voltages from equation (5.29) gives,

$$V_R = \frac{kT}{e}\ln\left(\frac{i_4}{i_3}\right) = \frac{kT}{e}\ln 8 \tag{5.34}$$

As the current flowing through R is $i/2$, the current is directly proportional to the absolute temperature,

$$i = \frac{2kT}{eR}\ln 8 \tag{5.35}$$

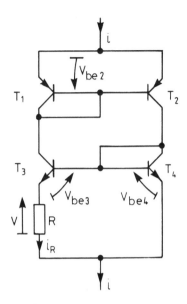

Figure 5.21 Circuit for an integrated silicon thermotransistor.

A modified version of this circuit is exploited in the commercially-available Analog Devices AD590 temperature sensor which provides a high-impedance current source of 1 µA/K with a supply voltage between 4 and 30 V DC. Its output is calibrated by the laser-trimming of internal resistors to (298.2 ± 2.5) µA for a temperature of 298.2 K. This type of sensor is particularly useful because it requires no complex external circuitry (e.g. a 3-wire bridge) and, as a current based detector, it can be used in remote applications where lead resistances would degrade the accuracy of detectors based on voltage readings.

5.7 Other Electrical Microsensors

5.7.1 Thermoswitches

A temperature switch, or *thermoswitch*, may be thought of as a thermal sensor with a discrete function that modulates a signal (i.e. on/off). Thermoswitches are often used in domestic and other appliances (e.g. an automatic kettle) as a control mechanism rather than as a measuring device. The most commonly used thermoswitch is the bimetallic switch which consists of two mechanically joined metal strips with differing thermal coefficients of expansion. The differential stress induced by a temperature change causes the strip to bend thus actuating a control function. Micromechanical equivalents of the bimetallic strip have been fabricated but solid-state versions are more common. These devices are in effect thermistors with a sharp change in resistance from a nominal value of, for example, 100 kΩ at 57°C or 75°C down to 100 Ω for only a 10°C change. Appropriate design ensures that resistance changes are small outside this temperature range.   Encapsulation of the microswitch in, for example, a TO-18 package can provide a low cost device for protection against over-heating.

5.7.2 Microcalorimeters

Calorimeters measure the amount of energy released or absorbed by an object through its associated change in temperature.   The temperature is usually measured using a thermistor or thermopile.   The mechanism by which heat is generated or removed can be associated with a chemical or biochemical reaction (§9.6) that results in a change in the free energy of the system.[3]   Thus it is necessary to have an active film as part of the microcalorimeter.   Consequently, microcalorimeters are classified here as chemical or biochemical sensors and are described later in Chapter 9.

5.8 Non-electrical Thermal Microsensors

There are several other types of thermal sensors that are non-electrical but are of interest here.   In particular, it is possible to clarify some of these sensors as microsensors because they require either thin film technology or could be micromachined in silicon.   They are discussed briefly below for the sake of completeness, and possible future miniaturisation.

5.8.1 Thermometers

The traditional method of measuring temperature is to monitor the thermal expansion of a gas, liquid or solid.   Like a bimetallic strip, the length of a solid or liquid L changes with its temperature,

$$L = L_0(1 + \alpha_1 T + \beta_1 T^2)$$
(5.36)

where L_0 is the length at a reference temperature (e.g. 0°C).   The change in length of liquids is usually linear with temperature since α_1 is the predominant coefficient.

The mercury thermometer can be operated over a temperature range of -35°C to +510°C and has an expansion coefficient of 1.82×10^{-4} /K.   A good discussion of traditional thermometry may be found elsewhere [5.2].   It would be relatively straight forward to realise an expansion thermometer in silicon, although its read-out would not be as easy as electronic devices such as the p-n diode.

5.8.2 Temperature indicators and fibre optical sensing

Certain materials (organic crystals) have been grown which undergo a colour change when heated or cooled.   These materials can be used to coat the surface of an object to act as a thin film temperature indicator.   As the temperature-induced material change causes the composition of white light to change, they can be

---

[3] This can be a change in enthalpy, e.g. the heat of combustion or adsorption.

Table 5.7 Thermochromic paints.

Temperature range (°C)	Colour	Minimum thickness (µm)
58 to 82	Yellow	10
82 to 102	Green	10
98 to 117	White	10

regarded as modulating temperature sensors. The sensitivity of these thermochromic films is low and they only cover a small temperature range. For example, synthetic chiral nemetic liquid crystals have the specification shown in Table 5.7.

These paints respond reversibly with temperature but their long term performance can be degraded by sunlight as the organic molecules are damaged by the UV radiation. Consequently, their practical use is somewhat limited.

A better material to use than a semi-liquid crystal is a glass which is much more robust. Figure 5.22 shows a possible arrangement using an optical fibre system to transmit signals through a temperature-sensitive arm (passing through an oven) and a reference arm [5.11]. The principle involved is that the phase difference between the two arms is shifted by temperature changes and is related to the decay time of a fluorescent active material. Table 5.8 shows a comparison between the performance of different sensor materials and the techniques employed for measurement and light modulation [5.11].

At present, instruments based on optical sensors are both large and expensive. However the use of integrated LEDs, fibre optic channels and thin optical materials may well provide a more practical silicon microsystem in the future.

Table 5.8 Comparison of several optical temperature sensor systems [5.11].

Sensor material:	Nd glass	Alexandrite	Ruby	RG:Nd glass
Technique	Decay time	Decay time	Decay time	Referenced absorption
Temp. range	-50°C to 220°C	20°C to 150°C	20°C to 140°C	-60°C to 200°C
Max. temp.	~420°C	~1,900°C	~1,900°C	~400°C
Calibration graph	Approx. linear	Deviates from linear	Deviates from linear	Approx. linear
Gradient	1.3×10^{-7} s/K	4.5×10^{-7} s/K	7×10^{-6} s/K	8.2×10^{-3} s/K
Signal strength	Strong	Medium	Medium	Strong
Light source	Convenient LED	Laser source (wavelength low on absorption profile)		Convenient LED
Modulation of light source	Electronic, inexpensive	Electro-optic, expensive	Mechanical, inconvenient	Electronic, inexpensive
Resolution	1°C	~0.5°C	-	-
Accuracy	±2.5°C	≤1°C	±2°C	±3°C

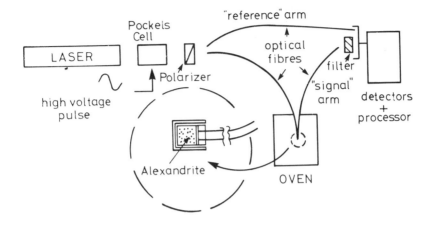

Figure 5.22 Fibre-optic temperature sensor [5.11].

5.8.3 Surface acoustic wave devices

Surface Acoustic Waves (SAWs) can be excited and made to travel through certain materials, such as $LiNbO_3$ or quartz. Figure 5.23 shows a schematic layout of a SAW delay line oscillator. The SAW is transmitted and received by the interdigital gold electrodes (100 nm) with a spacing of $\lambda/8$ in a $LiNbO_3$ substrate. The oscillation frequency is determined by the phase loop condition and is a function of temperature,

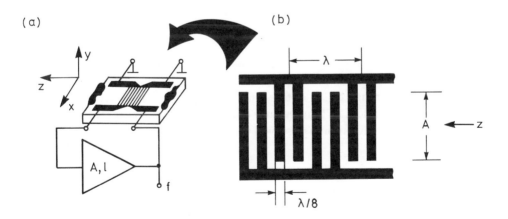

Figure 5.23 A SAW delay-line oscillator as a high resolution temperature microsensor [5.12].

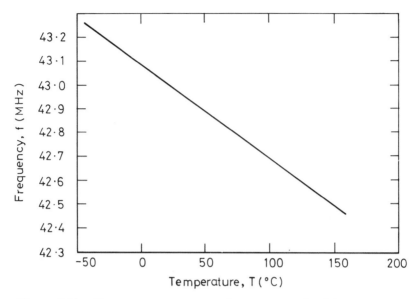

Figure 5.24   Frequency-temperature characteristic of a LiNbO$_3$ SAW temperature microsensor. From [5.12].

$$f(T) = \frac{(2\pi n - \Phi)}{\tau_d(T)} \qquad\qquad (5.37)$$

where Φ is the phase shift of the amplifier, τ_d the delay time of the SAW device and n the excited mode number. The delay time τ_d is temperature-dependent through the thermal expansion of the LiNbO$_3$ substrate and variation in propagation velocity. Suitable electronic design permits a single stable mode of oscillation f_0 (typically 40 MHz) which has a temperature dependence given by,

$$\frac{1}{f_0}\frac{df}{dT} \approx \alpha_l - \alpha_v \qquad\qquad (5.38)$$

where α_l and α_v are the thermal coefficients of expansion and velocity.

Figure 5.24 shows the response of a 43 MHz LiNbO$_3$ SAW delay-line oscillator over a temperature range of -50°C to +150°C [5.13]. The sensitivity is 4 kHz/°C and virtually independent of temperature over this range. The resolution is very high at 0.001°C but there is some observed fluctuation in signal of about ±0.2°C.

5.9 Performance Considerations and Commercial Devices

5.9.1 Relative performances of common thermal sensors

Thermal sensors form an important class of sensor because there are so many possible applications. The majority of these devices are used to measure the temperature rather than the heat flux or heat capacity of an object. The choice of the appropriate sensor to use depends upon the cost, size, sensitivity, temperature range, resolution etc. Consequently, it is only after a full consideration of the measurement or control problem that a suitable sensor may be determined.

Table 5.9 Relative performance of common electrical microsensors.

Type	Range (°C)	Resolution (°C)	Response time (s)	Output	Sensitivity (µV/°C)	Device	Comments
Thermocouple (Type T)	-160 to 370	±3	0.005	Linear, difference	~50	Thin foil	Rapid response foil construction. Need bridge to measure. Commercially available.
Thermopile (Type T)	0 to 70	±0.2	<1	Linear, difference	40	Micro	Research device. Very stable 0.0005 K/day.
Thermopile (n-Si/Al)	-200 to 100	±1	<1	~Linear, absolute	100	Silicon	Research device. High output. Can use α-Si on α-Ge instead of n-type Si.
Noise	-270 to 350	±2	$1/\Delta f$	Non-linear, absolute	~0.01	Resistor	Difficult to use and low sensitivity. Electronic circuit must be well designed.
PRT	-200 to 1,000	±0.1	~1	Linear, absolute	0.4 Ω/°C @ 100 Ω	Wire-wound	Commercial standard. Stable and precise but need bridge circuit.
Thermistor (oxide)	-270 to +450	±3%	5	Non-linear, absolute	Variable as NTC, PTC	Mini bead	Commercially available. Lower cost than PRT but non-linear.
Spreading resistor (n+-n-n+-Si)	-50 to +150	±1	10	~Linear, absolute	7 Ω/K @ 1 kΩ	Silicon	Cheap general purpose sensor with moderate accuracy. Integrated devices available.
Thermo-diode	-50 to +200	±1	10	Linear, absolute	-2 mV/K	Silicon	Cheap, general purpose sensor of moderate accuracy. Thermotransistors better characteristics (+TCR).
Thermo-transistor	-50 to 200	±1	10	Linear, absolute	1 µA/K @300 µA	Silicon	Low cost and with integrated IC are good general purpose temperature sensors.

Table 5.9 gives an overview of the relative performance and properties of some of the most common types of thermal microsensors available. The typical working temperature range of each microsensor is shown, together with its resolution, response time, nature of output, sensitivity, and type of device, e.g. thin film, miniature bead, or silicon microsensor. By using this table it should be

possible to select a candidate microsensor type for almost any application of interest.

5.9.2 Commercial thermal microsensors

There are a variety of thermal microsensors currently commercially available. Some examples are discussed briefly below with their properties summarised in Table 5.10. The cost is relative to the AD590JH which is £4.20 in 1993.

Figure 5.25(a) shows a photograph of a thin foil US ASA Type T thermocouple. The copper-constantan thermocouple foil is 50 μm thick and so the response is ultra-fast at a τ of 5 ms (grounded) and 10 ms (ungrounded). The low thermal mass also ensures that the contacted object does not have its temperature lowered significantly. The thermocouple has a good temperature range of *ca.* -160°C to +370°C and accuracy of ±1°C. This type of thermal microsensor is more expensive than others due to material costs. A second example, Figure 5.25(b), shows a photograph of a platinum film thermal resistor (manufactured in UK). A platinum ink is screen printed onto an alumina tile, fired and laser trimmed to give a resistance of (100 ±0.1) Ω at 0°C. The precise and stable element is made to BS 1904 Grade II and DIN 43760. The temperature range is -50°C to +150°C with a sensitivity of 0.385 Ω/°C. A typical operating resistance tolerance is ±0.5°C or 0.2 Ω. A 3-wire compensation bridge with $R_1=R_2=4$ kΩ and supply voltage of 12 V, gives an output of about 1 mV/°C. A platinum film resistance is more robust than the thin foil thermocouple and about one-third its price. Figure 5.25(c) to (e) show photographs of three low cost thermistors. Figure 5.25(c) shows a conventional Siemens NTC thermistor for industrial application.

Table 5.10 Some common commercial temperature microsensors.

Device	Type	Range (°C)	Error (°C)	Output	τ (s)	Cost factor
RS Components	Thin foil (Type T) thermocouple	-160,+370	±1	46 μV/K	5 ms	4.17
RS Components	Pt resistor	-50, +150	±0.5	0.385 Ω/K	~10	1.29
B572345S100M, Siemens	NTC Thermistor	-30, +125	3% on β	β const. 3,060 K	30	0.11
GM102W, Bowthorpe	NTC Thermistor	-80, +125	3% on β	β const. 2,910 K	5	0.95
Q63100-P371-D901, Siemens	PTC Thermistor	-25, +155	3% on β	100 Ω @ 25°C	~25	0.28
LM135H, Nat. Semi.	Thermodiode	-55, +150	±1	10 mV/°C	80	0.52
AD590JH, Analog Devices	Thermotransistor	-55, +150	±5	10 mV/°C	~50	1.00
LM35DZ, LM35CZ, Nat. Semi.	Thermotransistor	0, +100 -40, +110	±0.9 ±0.4	10 mV/°C 10 mV/°C	~50 ~50	0.50 0.88

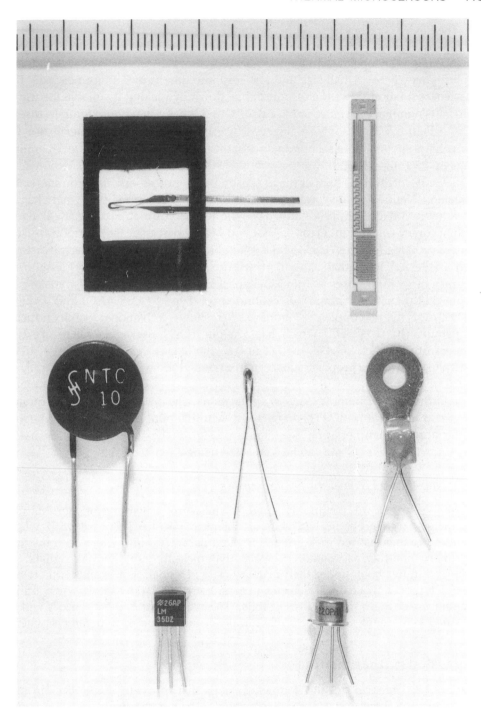

Figure 5.25 Commercial temperature microsensors. From left to right, and top to bottom: (a) Thin foil thermocouple, (b) platinum film resistor, (c) conventional NTC thermistor, (d) miniature bead NTC thermistor, (e) PTC thermistor, (f) thermotransistor IC and (g) constant current IC.

The device is rather large with a nominal resistance of (10 ±2) Ω and maximum continuous current of 5 A. A miniature version is shown in Figure 5.25(c) which is an NTC gas-encapsulated thermistor (Bowthorpe, GM102W) suitable for applications where a small size and fast response time (τ of 5 s) are needed. The glass encapsulation slows the response time down by making it suitable for use in fluids. Its nominal resistance is 1 kΩ at 20°C and it has a maximum temperature of 125°C. Figure 5.25(e) shows a Siemens (Q63100-P371-D901) PTC thermistor for low voltage circuits. It has a nominal resistance of 100 Ω at 25°C with an operating range of -25°C to +155°C.

Several integrated temperature sensors and processors are commercially available. National Semiconductor LM135H is a thermodiode that has a breakdown voltage proportional to absolute temperature at +10 mV/K. It has a good accuracy (maximum error of 1°C) and higher cost than the thermotransistor devices. Analog Devices (USA) make several temperature ICs (thermotransistors) such as the AD590JH that has a temperature range of -55°C to +150°C with an accuracy of ±5°C and uses a 3 pin TO-52 package. Analog Devices also make two 14 pin d.i.l. monolithic amplifiers calibrated for Type J (AD594AQ) and Type K (AD595AQ) thermocouples. Figure 5.25(f) shows a National Semiconductor temperature IC (LM35DZ) which has a 3 pin TO-92 package. There are two temperature ranges available: the DZ at 0 to 100°C and the CZ at -40 to +110°C. The voltage output is proportional to degrees centigrade with a typical accuracy of ±0.2°C.

For the use of remote sensor applications, the 590kH is a two terminal integrated temperature IC (TO-46 package) with the output current (in μA) equal to the absolute temperature, see Figure 5.25(g). The low cost device has a maximum error of 2 °C and can be connected using a twisted cable.

5.9.3 Non-thermal temperature microsensors

Thermal sensors are commonly used to measure a variety of non-thermal parameters, e.g. mechanical (§7.4.2) or chemical (§9.6). The ease with which temperature is measured makes it an excellent parameter into which to transduce a primary non-thermal signal. Sensors that measure non-thermal signals are discussed in this book under their primary class of signal. For example, there are several types of mechanical sensors (e.g. pressure and flow-rate) which utilise temperature changes in fine wires or foils. To save repetition these sensors will be classified as mechanical, magnetic, chemical etc. and discussed in detail elsewhere.

Suggested further reading

Readers are referred to the following texts that provide some background information on thermal sensors and associated instrumentation:

Noltingk BE (ed.): *Instrumentation Reference Book* (1988). Published by Butterworths & Co. Ltd, London. ISBN 0-408-01562-4. (620 pages. A comprehensive review of temperature and other sensors)

Rose-Innes AC: *Low Temperature Laboratory Techniques* (1973). Published by English Universities Press Ltd, UK. ISBN 0-340-17143X. (255 pages. A good discussion of low temperature sensing techniques)

References

5.1   Noltingk BE (ed.) (1988) *Instrumentation Reference Book*. Butterworths & Co. (Publishers) Ltd, London.

5.2   Norton HN (1989) *Handbook of Transducers*, Prentice-Hall International, London, 554 pp.

5.3   Brunetti L, Monticore E and Gervino G (1991) Thin-film thermopiles in microcalorimeters. *Sensors and Actuators A*, **25-27**, 633-636.

5.4   van Herwaarden AW (1986) Thermal sensors based on the Seebeck effect, *Sensors and Actuators*, **10**, 321-346.

5.5   van Herwaarden AW, van Duyn DC, van Oudhesden BW and Sarro PM (1989) Integrated thermopile sensors. *Sensors and Actuators A*, **21-23**, 621-630.

5.6   Middelhoek S and Audet SA (1989) *Silicon Sensors*, Academic Press Ltd, London, p.163.

5.7   van Herwaarden AW (1984) The Seebeck effect in silicon ICs. *Sensors and Actuators*, **6**, 245-254.

5.8   Urban G, Jachimowicz A, Kohl F, Kuttner H, Olcaytug F and Kamper H (1990) High-resolution thin film temperature sensor arrays for medical application. *Sensors and Actuators A*, **21-23**, 650-654.

5.9   Wolf HF (1969) *Silicon Semiconductor Data*, Pergamon Press, Oxford, UK.

5.10   McNamara AG (1962) Semiconductor diodes and transistor as electrical thermometers. *Rev. Sci. Instrum.*, **33**, 330-333.

5.11   Grattan KTV and Palmer AW (1987) Fibre-optic-addressed temperature transducers using solid-state fluorescent materials, *Sensors and Actuators*, **12**, 375-387.

5.12   Neumeister J, Thum R and Lüder E (1990) A SAW delay-line oscillator as a high-resolution temperature sensor. *Sensors and Actuators A*, **21-23**, 670-672.

Problems

5.1   State, giving reasons, the type of thermal sensor that you would use to measure the temperature of (a) a stationary metal heating block used to seal plastic boxes on a conveyor belt, (b) a small electronic component soldered onto a stationary PCB and (c) a liquid flowing down a 10 mm pipe.

5.2   What are the possible advantages and disadvantages of using a thermoresistor in a real application?

5.3   State the working temperature range of a thermodiode. Explain the underlying principles that result in its limited temperature range.

5.4   Produce a simple physical model of a glass bead thermal microsensor sensing the temperature of still air. Give the equivalent electrical circuit of your model. Sketch out the dynamic response of your model to a rectangular pulse of heat generated by a combustion process lasting 5 seconds.

6. Radiation Microsensors

Objectives

- [ ] To introduce the subject of radiation microsensors
- [ ] To review their operating principles
- [ ] To present some typical applications

6.1 Basic Considerations and Definitions

6.1.1 Nuclear and electromagnetic radiation

Radiation is the emission of either particles or electromagnetic rays from a source. Particles emitted from an object are usually nuclear particles, i.e. particles which can be emitted from a nucleus. These may be generated by the decay of a radioactive material or by the interaction of a nucleus with another energy source. Nuclear particles, unlike electromagnetic rays, have a finite rest mass. The mass is used to classify nuclear particles into several groups: baryons (heavy particles), mesons (medium particles) and leptons (light particles).

Alpha- (α-) and beta- (β-) rays are the names of two common nuclear particles which are found to be emitted from naturally occurring materials. An α-particle is just the charged nucleus of a helium atom, i.e. a helium ion. For example, the substance radium is often found in granite and is naturally radioactive. Radium (Ra) has an atomic mass A_Z of 226 (and atomic number Z_A of 88) and decays to produce radon gas (A_Z=222, Z_A=86) and an α-particle (A_Z=4, Z_A=2). This process can be written as,

$$^{226}_{88}\text{Ra} \rightarrow\,^{222}_{86}\text{Rn} +\,^{4}_{2}\text{He} \tag{6.1}$$

The emitted α-particle consists of two protons and two neutrons (charge 2+) and has a kinetic energy of between 4 and 5 MeV. Radon, which if breathed in is a hazard to our health, is also radioactive and can decay to produce another α-particle of 5.5 MeV. As radon is a heavy gas it concentrates near the ground, so pumps can be installed at floor level to remove the radioactive gas from houses built on granite. Other radioactive materials can decay to emit neutrons and protons which also belong to the class of baryons. Neutrons and protons have similar masses (≈0.25 that of an α-particle) but the proton has a single positive charge.

β-particles are energetic particles with the rest mass of an electron that have been emitted by a source (radioactive or otherwise) and have a much lower mass, thus they are called leptons. β-particles are electrons but a positively charged particle version, β⁺, of the same weight is called a positron. Table 6.1 summarises the common nuclear particles that need to be detected by radiation sensors. A near massless particle called the neutrino is often produced in radioactivity. The mass of the neutrino is unknown but very small and it can have very high energies.

Electromagnetic radiation is in essence a flux of massless particles or quanta of energy. Electromagnetic "particles" have zero rest mass energy, and are usually described as waves rather than as particles. They all travel at the speed of light v_c in vacuum and have a mass equivalent energy E_R which relates to their frequency f or wavelength λ, by

$$E_R = hf = \frac{hv_c}{\lambda} \qquad\qquad (6.2)$$

where h is Planck's constant and has a value of 6.63×10^{-34} Js.

As the energy of an electromagnetic particle is proportional to its frequency, it

Table 6.1 Common particles making up nuclear radiation.

Name	Symbol	Mass (MeV)	Charge
Baryons:			
α-particle	He, α	3,755.8	+2e
Deuteron	d	1,875.6	0
Proton	p	938.3	+e
Neutron	n	939.6	0
Mesons:			
Pion	π^+, π^0, π^-	139.6 or 135.0	+e, 0, -e
Kaon	K^+, K^0, K^-	493.8 or 497.8	+e, 0, -e
Leptons:			
β⁺-particle	β^+, e^+	0.51	+e
β⁻-particle	β^-, e^-	0.51	-e
Neutrino	0	>0	0

is usual to classify electromagnetic radiation according to its frequency as a fundamental property.

Figure 6.1 shows the entire electromagnetic spectrum from the highest energy cosmic ray particles ($E_R \sim$ GeV) down to the lowest energy radio waves ($E_R \sim$ neV). Many of these types of radiation are commonly used in our everyday lives. X-rays are used to image our internal bone structure in medicine. Ultra-violet radiation is used to produce an artificial sun-tan. Visible radiation (normally called light) is detected by the human eye but it occupies only a small range in wavelength (400 to 700 nm). However, visible radiation is of major importance in many applications (cameras, opto-electronics etc.). The utilisation of radiation within other types of sensors is common.

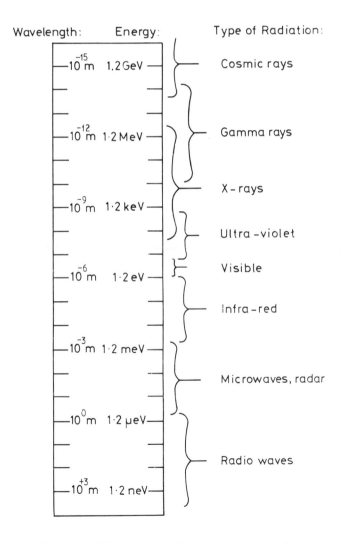

Figure 6.1 The spectrum of electromagnetic radiation.

Infra-red radiation can be used to probe the properties of molecules or measure temperature (§5.1).   Microwaves and radar radiation are used in proximity sensors (§7.3.5), whereas microwave radiation can also be used for heating purposes. Finally, radio waves are used in many communication systems.

The devices needed to detect the radiation itself are discussed in this chapter. The principles of the basic devices will be discussed here which may then be exploited in other sensor systems.

6.1.2 Radiation measurands

Several parameters are used to characterise the properties of nuclear particle radiation. First, the energy of the particle relates to the speed at which it travels. For non-relativistic particles the kinetic energy E_k is related to the particle velocity v_R by,

$$E_k = \frac{1}{2} m_R v_R^2 \tag{6.3}$$

where m_R is the mass of the radiation particle. Next, it is important to know the flux of particles Φ_R emitted per second. The power P is the rate of change of energy per unit second and is related to the kinetic energy by,

$$P = \frac{dE_k}{dt} = E_k \Phi_R \tag{6.4}$$

When a beam of particles is emitted from a source it is often important to know the power per unit area A or solid angle Ω.

Table 6.2 summarises some basic radiometric (electromagnetic) and photometric (visible light) terms together with those for particles.

Table 6.2 Some important radiation measurands with SI units.

Term	Particle	Radiometric	Photometric
Energy	Kinetic energy (J)	Radiant energy (J)	Luminous energy (lm s)
Power	Particle flux (s^{-1})	Radiant flux (W)	Luminious flux (lm)
Intensity (planar)	Particle density (s^{-1} m^{-2})	Radiant exitance (W/m^2)	Luminous intensity (lm/m^2)
Intensity (angular)	Particle intensity (s^{-1} sr)	Radiant intensity (W/sr)	Luminous intensity (cd)

A radiation source usually emits radiation in a random direction. So a small source can be regarded as emitting at a certain power level into a solid angle of 4π steradians. The radiant intensity I_R is thus given by

$$I_R = (P/\Omega) = (P/4\pi) \quad \text{(units of W/sr)} \tag{6.5}$$

For collimated sources of radiation, it is possible to use the planar intensity which is simply the power per unit area.

$$I_R = P/A \tag{6.6}$$

At a large distance from a point radiation source (e.g. the Sun), the radiation is approximately collimated and so the angular variation may be neglected.

6.1.3 Classification of radiation sensors

Radiation sensors can be classified as non-contacting sensors because they detect remotely the emission of various particles or electromagnetic radiation, e.g. UV, IR

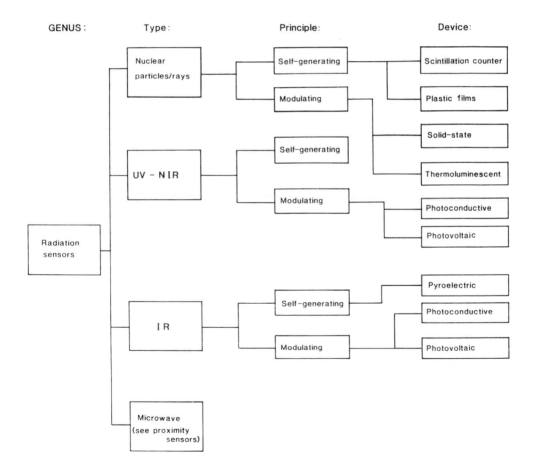

Figure 6.2 Classification scheme of radiation sensors.

(see Figure 6.2). Radiation sensors can be conveniently subdivided into two classes: nuclear particle and electromagnetic radiation. Nuclear particle radiation sensors are frequently used to measure the particles or substances emitted by a radioactive substance, such as α-particles, β-particles (fast electrons), and heavy particles (neutrons, protons, etc.). Nuclear radiation sensors can measure electromagnetic radiation such as γ-rays and X-rays emitted by radioactive sources. Nuclear radiation sensors that detect both particles and electromagnetic radiation are sometimes referred to as nucleonic detectors.

There are a variety of nuclear radiation sensors that are not regarded here as microsensors. These include the gas detectors which generally consist of a gas-filled chamber. As the nuclear particle travels through the gas it collides with the molecules and forms an ion pair. These can be visualised in a cloud chamber or the resulting ions are accelerated by an electric field and multiplied. Different methods are employed to measure either the current (ionisation chambers) or individual arrival of pulses (proportional counters and Geiger counters). Under certain conditions the movement of a charged particle in a transparent medium can also be detected by the characteristic Cherenkov light emitted. These so-called Cherenkov detectors, as well as the other types of macrosensors, are described in standard books on nuclear physics, e.g. [6.1].

Most of the electromagnetic spectrum can be measured using a solid-state detector such as a silicon photodiode (X-ray to NIR) or a pyroelectric detector (IR). These microsensors measure the radiant energy or flux at a single point in space, and so can be thought of as "point" or "discrete" sensors. This type of sensor is often used as a principle for other kinds of sensor. For example, a photodiode or pyroelectric sensor is used in a mechanical proximity (non-contacting) sensor. These secondary applications of radiation sensors will be described in Chapter 7, e.g. mechanical microsensors.

Finally, the use of microelectronic technology (Chapters 3 and 4) permits the low-cost production of 1-d and 2-d arrays of radiation microsensors. A discussion of such devices as photodiode arrays, MOS linear image sensors, CCD and CID cameras is in Chapter 12 on microsensor array devices.

6.2 Nuclear Radiation Microsensors

6.2.1 *Scintillation counters*

The scintillation counter consists of an active material which converts the incident nuclear radiation to pulses of light, a light-electrical pulse converter (e.g. a photomultiplier tube) and an electronic amplifier/processor. The active material that scintillates is an inorganic or organic crystal, a plastic *fluor* or a liquid. The size of the scintillator varies enormously depending on the energy of the radiation, from thin solid films to large tanks of liquid to measure cosmic rays. Consequently, some inorganic and plastic scintillators may be classified as microsensors although these generally detect softer electromagnetic radiation. The associated signal processor can be large (e.g. a photomultiplier tube) but miniature systems have been developed for application in space vehicles and satellites.

Table 6.3 Some properties of inorganic scintillator materials.

Material	Detects	Density (g/cm^3)	Light output[1] (%)	Decay time (μs)	Comments
NaI(Tl)	γ-, X-	3.67	230	0.23	Good γ-ray, X-ray detector. Widely used. Hygroscopic and shock dependent.
NaI(Pure)	γ-, X-	3.67	440	0.06	Operated at 77 K.
CsI(Tl)	X-	4.51	95	1.1	Used in satellites as X-ray detector (not hygroscopic). Also good α-ray detector.
$Bi_4Ge_3O_{12}$	X-	7.13	18	0.3	Used in tomographic scanners to detect 170 keV X-rays.
CsF	X-	4.64	41	0.005	Used in tomographic scanners as alternative to $Bi_4Ge_3O_{12}$.
LiI(Eu)	n	4.06	75	1.2	Neutron detector but expensive and hygroscopic.
CaF_2(Eu)	β	3.17	110	1.0	Not hygroscopic so suitable as β-detectors.

[1] Expressed as 1% of anthracene crystal.

Table 6.3 provides a list of some common inorganic scintillator materials together with their density, light output, and decay times. The selection of a scintillator depends upon the application requirements; they all produce a pulse height proportional to the energy E_R of the radiation absorbed and a count-rate that depends upon the radiation flux Φ_R:

$$E_R \propto \text{pulse height}; \quad \Phi_R \propto \text{count-rate}. \qquad (6.7)$$

Sodium iodide (with a thallium additive) is commonly used in X-ray and γ-ray detectors but it is a very hygroscopic material. Thallium doped NaI has a rather long decay constant (230 ns) although this can be reduced by omitting the dopant and running at a low temperature (see Table 6.3). CsI is also used because it is more resistant to shock and less hygroscopic. More importantly, it has a larger adsorption coefficient and can be used in thin layers. Thus, CsI(Tl) scintillation detectors have been used in space vehicles and satellites. CsI(Tl) radiation sensors are also used as α-particle detectors because they are not moisture-sensitive.

$Bi_4Ge_{13}O_{12}$ has a high absorption of 170 keV X-rays and was developed specifically for tomographic scanners. Its efficiency is lower than NaI(Tl) and it has a long decay time. An alternative material is CsF which has a greater efficiency (~ factor of 2) and has a very short decay time of about 5 ns.

LiI(Eu) is used to detect neutrons by using enriched ^{6}Li rather than ^{7}Li. At room temperature it measures slow neutrons but at very low temperatures (e.g. *ca.*

77 K) it can measure fast neutrons. However LiI(Eu) detectors are expensive and very hygroscopic.

CaF$_2$(Eu) can be used as a β-particle sensor because it is inert, and is not hygroscopic. It is also interesting to note that ZnO (doped with gallium) and ZnS (doped with silver) can also be used as scintillation detectors. These materials are important as they are used in other microsensors.

Organic scintillator materials are used as an alternative to inorganic materials. Anthracene crystals were among the first to be used but now a variety of plastics are made. These plastics are easier to process than inorganic materials and have very small decay times (~ns). Consequently, they are widely used to measure nuclear radiation, see Table 6.4.

Table 6.4 Properties of some commercially available organic scintillator materials.

Material code	Detects	Density (g/cm^3)	Light output (%)	Decay time (ns)	Comments
NE102A	γ, α, β, fast n	1.032	65	2.4	General purpose
NE104	γ, α, β, fast n	1.032	68	1.9	Ultra-fast counting
NE114	γ, α, β, fast n	1.032	50	4.0	Cheaper for large arrays

The main disadvantage of using plastic scintillation sensors is that they soften at moderate temperatures (~75°C). However, their general inertness and processability makes them commercially viable. Plastic scintillator materials are only capable of producing a Compton scattering of the incident radiation due to their low atomic number. In contrast, inorganic scintillator materials have other interactive processes such as the photoelectric effect and pair production. Photoelectric effect is dominant for low energy radiation as discussed in the next section.

6.2.2 Solid-state detectors

The use of a solid-state or semiconductor material in a nuclear radiation sensor is highly desirable. The majority of interest has centred upon the use of silicon or germanium although other semiconductors such as CdTe, HgI$_2$ and GaAs have also been studied.

Radiation is absorbed by semiconducting materials and the level of adsorption varies with the material and radiation energy. However, as stated above, there are three major processes that lead to the absorption of radiation: at low energies the photoelectric effect dominates, at intermediate energies the Compton effect dominates and at high energies pair production dominates, see Figure 6.3 [6.1].

Figure 6.4 shows the variation of the absorption coefficient α of several semiconductor materials with photon energy. The contributions of the photoelectric effect, Compton scattering and pair production are shown separately [6.2].

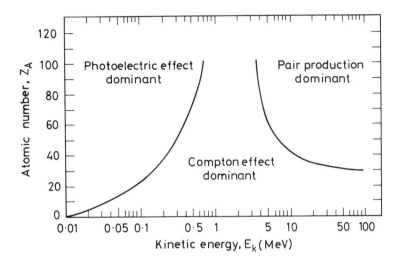

Figure 6.3 Interaction of X-rays and γ-rays with matter.

The absorption coefficient is simply a measure of the fractional decrease in incident radiation intensity I_{R0} with distance into the material, and is therefore exponential in profile,

$$I_R(x) = I_{R0} \exp(-\alpha x) \quad \text{or} \quad \alpha = \frac{-dI_R}{I_R dx} \qquad (6.8)$$

In scintillation sensors, the absorbed radiation generates photons which are in turn detected. However, in solid-state detectors the absorption leads to the generation of conduction electrons and thus an electrical signal.

The photoelectric effect dominates at low energies (below about 100 keV) and is important in detecting X-rays and soft gamma-rays. The phenomenon, which was first explained by Einstein, involves a photon with a quantum of energy hf transferring all its energy to electrons bound to atoms in the lattice (i.e. valence electrons). The excited electron has kinetic energy E_k where

$$E_k = (hf - E_b) \qquad (6.9)$$

The binding energy of the valence electron E_b varies with the choice of material and shell. The excess of electron-hole pairs leads to an increase in the bulk conductivity.

At intermediate energies (0.1 to 1 MeV) the Compton effect dominates whereby the radiant photon only passes on part of its energy to the valence

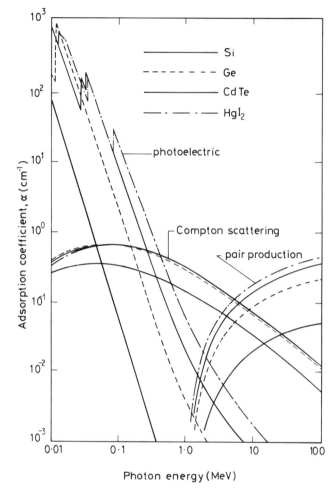

Figure 6.4 Absorption coefficient of semiconductor materials used in nuclear radiation sensors. From [6.2] (© 1973 IEEE).

electron.   The lower energy electron can then be fully absorbed by the photoelectric effect, Compton scattered again or escape altogether. The Compton effect leads to conduction electrons (and lower energy electrons detected by scintillation detectors), which once again modify the bulk conductivity.

Finally, at photon energies above 1.02 MeV the incident photon can be adsorbed to produce an electron-positron pair. The rest mass energy required to create the pair (1.02 MeV) is lost and any excess energy goes into kinetic energy. The resultant electrons and holes lose their energy through collisions (e.g. Compton) with the positron finally combining with an electron to produce two gamma rays each of 0.51 MeV. Again these gamma rays can interact by Compton or photoelectric effect to produce more electron-hole pairs.

Although these three effects combine to produce electron-hole pairs, the efficiency of this process is rather low. There are two reasons for this: first the

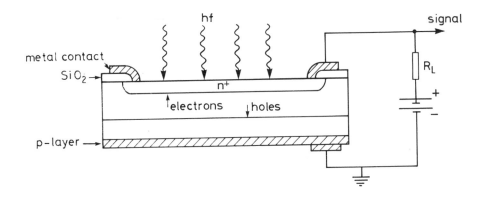

Figure 6.5 Schematic diagram of a *p-n* photodiode.

quantum efficiency η can be low, that is the number of electron-hole pairs produced by each incident photon is low. The quantum efficiency is principally determined by the absorption coefficient (see Figure 6.4), but falls off dramatically at higher energies. Secondly, the conductivity of silicon is high compared to the photocurrent generated and so measurement is difficult. Consequently, solid-state nuclear radiation sensors are photodiodes rather than photoconductors. The photodiode is basically a *p-n* junction under reverse bias, see Figure 6.5. The photon generates an electron-hole pair in the depletion region which are then separated out and so an electric current flows.

The resolution of solid-state photodiodes exceeds that observed for NaI(Tl) detectors by a factor of about ten. Consequently, solid-state photodiodes can be used to resolve energy spectra to a much higher degree of accuracy. Solid-state detectors can also be used in a vacuum and at low temperatures (e.g. 77 K) to improve efficiency. However, large detector areas are expensive and scintillation counters have a much faster response time which reduces the counting errors.

6.2.3 *Other nuclear radiation microsensors*

The total amount or dose of radiation falling upon a surface may be measured using a thin (5 μm) plastic polycarbonate film. The film is mechanically damaged as highly ionising particles (e.g. α-particles) travel though. Subsequent etching of the film reveals tracks which can be observed through a microscope and counted. Alternatively, the dose can be measured using a thermoluminescent material, e.g. LiF. The incident radiation excites metastable energy states in the material. The subsequent heating of the material produces visible photons as the states are accessed. Thus, the films are reusable and are commonly used in dosimetry to measure levels of X-rays, β-rays and γ-rays (30 keV to 2 MeV).

6.3 Ultra-violet, Visible and Near Infra-red Radiation Microsensors

6.3.1 General sensor selection by wavelength

Important regions of the electromagnetic spectrum for microsensors are the ultra-violet (UV, 0.002 to 0.4 μm), the visible (Vis, 0.4 to 0.7 μm) and the infra-red (IR, 0.7 to 500 μm). The infra-red region may be further subdivided into the near infra-red (NIR) region of about 0.7 to 1.7 μm and the main infra-red region of about 1.7 μm to 500 μm. It is useful to examine the UV-Vis-NIR region as a whole because the common semiconducting materials operate over this region. Figure 6.6 shows the operating range of sensors employing semiconductor materials, in the UV, Vis, NIR and IR regions. Clearly, silicon is an important material as it covers a wide part of the UV-Vis-NIR spectrum. Thus, silicon microelectronic devices such as Si photodiodes are commonly used to detect radiation and will be discussed in detail later in this chapter. Other materials and principles are used to detect IR radiation, such as PbS cells or pyroelectric detectors, and these are described in the next section.

6.3.2 Photoconductive cells

Photoconductive cells are semiconductor sensors that utilise the photoconductive effect in which light striking the photoconductive material reduces its resistance. They are sometimes referred to as light dependent resistors. Figure 6.7 shows the basic configuration of a photoconductive cell. In darkness, the cell resistance is

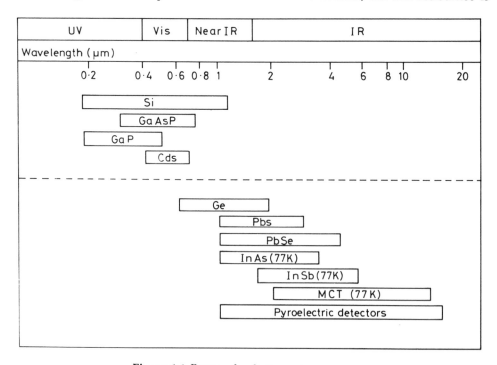

Figure 6.6 Range of radiation microsensors.

very high (typically ~ MΩ) due to a band-gap which is large compared to kT. The resistance falls as the intensity of light shining onto it is increased.

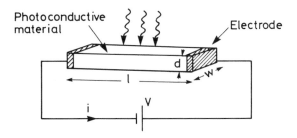

Figure 6.7 Basic structure of a photoconductive cell.

Cadmium sulphide is widely used as a photoconductive material. Its spectral range is essentially the visible region from 400 nm to 800 nm (10% of peak value) with a peak response at about 550 nm, see Figure 6.6. Crystalline CdS has a band gap of 2.41 eV with donor and acceptor levels that are nearly fully occupied. A photon will excite a valence electron into the conduction band creating an electron-hole pair that can contribute to the conduction process. The acceptor level near the valence band can capture the free holes easily and thus reduce the recombination probability. The change in electrical conductivity $\Delta\sigma$ is given by the following equation:

$$\Delta\sigma = eN_t(\mu_n\tau_n + \mu_p\tau_p) \tag{6.10}$$

where μ_n and μ_p are the mobilities of the free electrons and holes, τ_n and τ_p are the life-times of the electrons and holes and N_t is the number of carriers generated per second per unit volume. For a CdS cell, the life-time of the holes is short so it behaves like an n-type semiconductor where

$$\Delta\sigma \approx eN_t\,\mu_n\tau_n \tag{6.11}$$

The number of electrons that flow between the electrodes due to excitation by one photon (i.e. gain A) is related to the transit time t_t between the electrodes ($l^2/V\mu_n$):

$$A = \tau_n/t_t = \mu_n\tau_n V/l^2 \tag{6.12}$$

For a typical CdS cell, the mobility μ_n is 300 cm^2/V s, τ_n is 1 ms, l is 0.2 mm and V is 1.2 V, then the gain is nearly a factor of 1,000. Thus, there is a high multiplication that leads to a sensitive device.

From Ohm's law, the change in device resistance ΔR is related to the change in conductance (see Figure 6.7) by

$$\Delta R = \frac{1}{\Delta \sigma \, d}\left(\frac{l}{w}\right)$$

(6.13)

where d is the thickness, and (l/w) the aspect ratio of the device. Thus, meandering electrode patterns are used to increase the sensitivity of the cell. The spectral response of a CdS cell is extended by adding CdSe. However, Cd(S:Se) and CdSe cells are often referred to as CdS cells for the sake of convenience.

Figure 6.8 shows the typical variation of the resistance with illuminance of a CdS photoconductive sensor. The dark resistance of ~1 MΩ at 0.1 lx falls nearly exponentially with log illuminance to about 1 kΩ at 1,000 lx. The rise time of photoconductance cells is long at ~50 ms but the decay time is slightly less at ~40 ms (at 10 lx).

The temperature characteristic of photoconductive cells is generally poor and depends upon the composition, fabrication method and light level. Furthermore, the baseline resistance drifts with time, perhaps by 20% over 1,000 h. For these reasons, the applications of photoconductive cells are limited more to the control

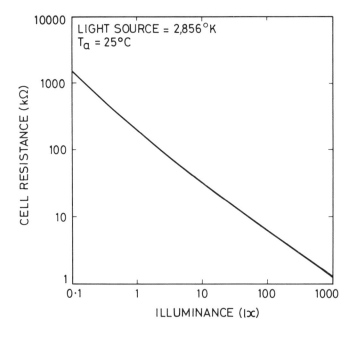

Figure 6.8 Resistance of a CdS photoconductive cell with illuminance. From [6.3].

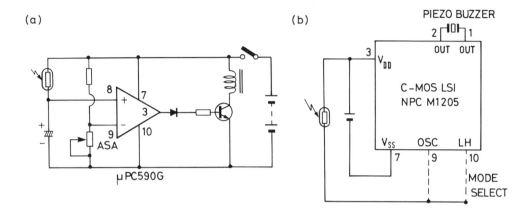

Figure 6.9 Two applications of a photoconductive cell: (a) in an electronic shutter, and (b) in a melody greeting card.

and detection of light rather than its absolute measurement. Some of the common applications are as:

- automatic dimmers;
- light switches;
- photoelectric servos;
- electronic shutters;
- melody greeting cards;
- photoelectric relays.

Figure 6.9 shows two basic operating circuits for the application of a photoconductive cell as a light-activated electronic shutter and in a greeting card which plays a tune when opened!

6.3.3 *Photodiodes*

Photodiodes may be classified as potentiometric radiation sensors because the radiation generates a voltage across a semiconductor junction. This phenomenon is known as the photovoltaic effect. Strictly speaking the photons generate free carriers (electron-hole pairs) which move in response to local fields due to a variation in doping (a semiconductor-semiconductor junction), to a variation in composition (a heterotransition) or to a variation in both (a heterojunction). The main types of photodiode are:

- *p-n* photodiode;
- *p-i-n* photodiode;
- Schottky-type photodiode;
- Avalanche photodiode.

Photodiodes are widely used to detect the presence, intensity and wavelength of UV-NIR radiation. The advantage of photodiodes over photoconductive cells are:

- a higher sensitivity;
- faster response time;
- smaller size;
- better stability;
- excellent linearity.

Photodiodes can be used to cover the entire UV-NIR spectral range through the choice of material, device structure and external optical filters. Figure 6.10 shows the radiant sensitivity[2] against the wavelength of six types of commercial photodiodes.

Silicon photodiodes can detect radiation from the UV to NIR (190 to 1,100 nm) with a peak at 960 nm. PIN (*p-i-n*) silicon diodes have a range of 320 to 1,100 nm. GaAsP Schottky-type photodiodes have a narrower spectral range of UV to visible

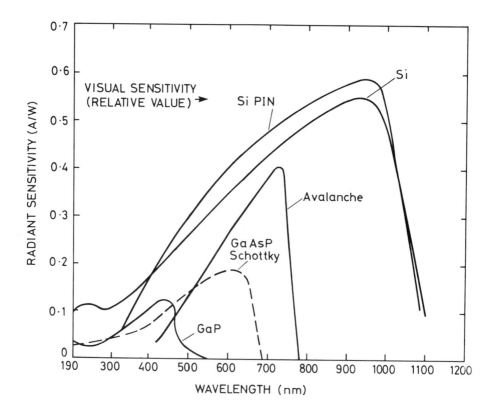

Figure 6.10 Spectral response of six photodiodes.

---

[2] This is the ratio of radiant power (in watts) on the device to photocurrent output.

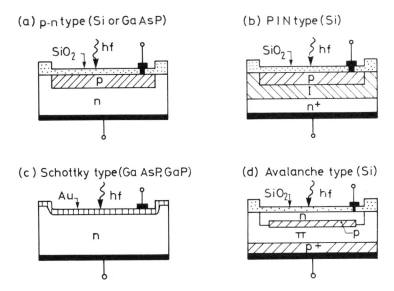

Figure 6.11 Structure of four types of photodiode.

(190 to 680 nm) as do GaP photodiodes (190 to 550 nm). Finally, silicon avalanche photodiodes cover the Vis-NIR range (400 to 800 nm).

Figure 6.11 shows the construction of the four main types of photodiode: the *p-n* type, PIN type, Schottky type and Avalanche type. The *p-n* photodiode consists of a *p-n* junction (e.g. Si) with a SiO_2 coating to help reduce the dark current and improve reliability. As described earlier, the absorption of a photon (photoelectric effect) creates an electron-hole pair which are driven by the junction field to the doped regions, see Figure 6.12(a). This creates a photovoltage *V* as shown by Figure 6.12(b).

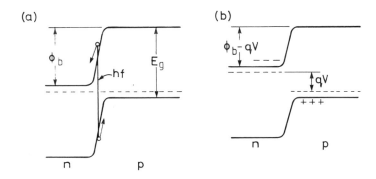

Figure 6.12 Generation of a photovoltage at a *p-n* junction [6.4].

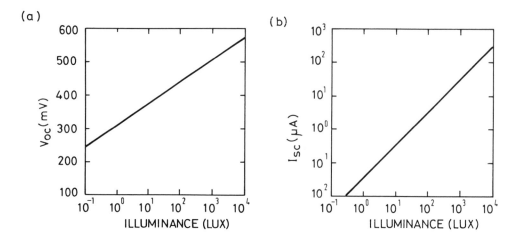

Figure 6.13 Output of a typical silicon photodiode against illuminance:
(a) open-circuit voltage and (b) short-circuit current at 25°C.

The open-circuit voltage V_{oc} of a photodiode can be measured when the external load resistor R_L is high and is given by,

$$V_{oc} = \frac{kT}{e} \ln\left(\frac{I_L}{I_S} + 1\right)$$

(6.14)

where I_L is the photocurrent (proportional to amount of light) and I_S is the photodiode reverse saturation current. Thus, the open circuit voltage is proportional to temperature and log illuminance. Figure 6.13 shows that V_{oc} is very linear over 4½ decades of light level. The short-circuit current I_{sc} is obtained when the load resistor is in effect zero so that no voltage is developed across the diode. This is also very linear with illuminance. Photodiodes are usually operated under reverse bias or in a virtual earth circuit. Figure 6.14 shows the two basic arrangements.

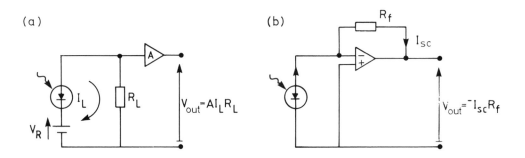

Figure 6.14 Two basic circuits for operating a photodiode: (a) reverse bias and (b) virtual earth.

The PIN type photodiode has a thin insulating layer between the p-type and n^+-type material. This means that the depletion region thickness can be modified to optimise the quantum efficiency and frequency response. The lower junction and package capacitance produces a much faster response than that for a typical p-n diode ($\sim$0.4 μs).

The Schottky type photodiode has an ultra-thin metal film ($\sim$100 Å) that forms a Schottky barrier with an n-type semiconducting material. The ultra-thin metal film enhances the sensitivity of the diode to the UV range where the adsorption coefficient in semiconducting materials is high. It is necessary to use an anti-reflection coating such as 500 Å of ZnS but over 95% of the incident radiation (λ is 633 nm) is transmitted into the silicon substrate.

The height of the Schottky barrier ϕ_B ideally depends upon both the work function of the metal ϕ_m and the work function of the semiconductor ϕ_s, where

$$\phi_B = (\phi_m - \phi_s) \qquad (6.15)$$

The current density J through the ideal rectifying contact is given by the well-known equation,

$$J = J_0 \exp\left(-\frac{\phi_B}{kT}\right)\exp\left(\frac{qV}{kT}-1\right) \qquad (6.16)$$

where J_0 is $\sim$120 T^2 A/cm^2 (when the effective electron mass is one) and V is the applied voltage. At UV wavelengths ($hf > E_g$) electron-hole pairs are generated inside the Schottky barrier and separated by the local field. The resultant change in barrier height can be measured as before. At longer wavelengths ($hf > \phi_B$), electrons within the metal are excited enough to cross the barrier into the semiconductor. However, the probability is lower than band to band excitation (at lower wavelengths) as seen from Figure 6.15. In practice, the barrier height of, for example, GaP does not vary according to equation (6.16) due to the presence of surface states.

With an n-type semiconductor, the barrier height seen by the metal is essentially independent of the bias voltage. The barrier height is now set by the difference between the conduction band edge E_c and Fermi level at the surface E_o,

$$E_c - E_o = \phi_B \approx \frac{2}{3}E_g \qquad (6.17)$$

Typically, $(E_c - E_o)$ is $\frac{2}{3}E_g$ for many materials in which the junction characteristic is surface-state limited. Choosing the semiconductor material, e.g. GaAsP or GaP, permits the low-energy threshold to be moved along the spectrum.

The Avalanche Photodiode (APD) is operated under a high reverse bias voltage in which the photon-generated carriers are excited to sufficient levels to collide with other atoms and produce secondary carriers. This process occurs repeatedly and is called the Avalanche effect. This internal multiplying

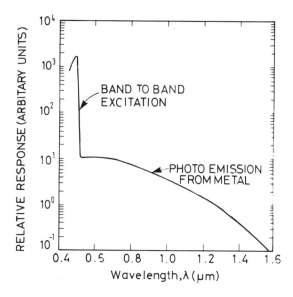

Figure 6.15 Spectral response of a Schottky-type photodiode showing the two characteristic regions.

mechanism leads to a sensitivity to low light levels in the Vis-NIR spectral region. Figure 6.16 shows the high quantum efficiency of a Si avalanche photodiode [6.5]. The condition for avalanche to occur is that the kinetic energy E_k of the electron exceeds $\frac{3}{2}E_g$ and creates another carrier,

$$E_k \geq \tfrac{3}{2} E_g \tag{6.18}$$

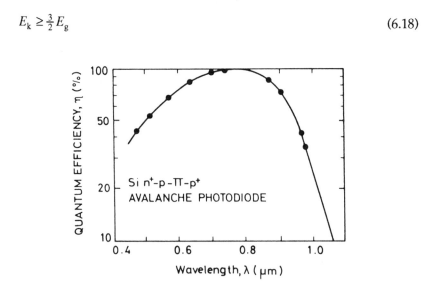

Figure 6.16 Quantum efficiency of a silicon avalanche photodiode [6.5]. (Reproduced by permission of John Wiley & Sons Inc., © 1985.)

Avalanche diodes also have ultra-fast response with cut-off frequencies up to the GHz region. Typical break down voltages of 100 V for Si APD are observed with photocurrents of up to 1 mA near breakdown.

6.3.4 Phototransistors

The light sensitivity can be improved through the use of a phototransistor rather than a photodiode. In addition, phototransistors can be used in applications where readers, counters or encoders are needed. The rise and fall times are typically 1 to 5 μs - although the use of a photodarlington configuration reduces this to 50 μs but improves sensitivity by a factor of 10 or so.

6.3.5 Relative performance of commercial UV-NIR radiation sensors

A wide variety of UV-NIR radiation sensors are commercially available. The devices come in a wide range of IC packages with optical accessories, e.g. wide or narrow angle lenses, optical filters etc. Table 6.5 summarises the relative performance of some commercial radiation UV-NIR sensors.

From Table 6.5, it is possible to see the wide range of devices available for any particular application. The sensitivity of the devices varies but is comparable with that of the human eye. At very low light levels, it is necessary to use scintillation counters or photomultipliers rather than silicon-based microsensors. However, the cost and size of these photonic devices is considerably higher.

The temperature sensitivity of photodiodes/phototransistors is lower than that of photoconductive cells. The shunt resistance (and hence dark current)

Table 6.5 Relative performance of some commercial UV-NIR radiation microsensors.

Device/ manufacturer	Size	Type	Range	Typical output (at 25°C)	Response (τ_r, τ_d)	Cost
MPY54C569 (Sentel)	4.3×5.1 mm	CdS	Vis	0-1,000 lx @ 20M-1kΩ	30 ms, 10 ms	Low
S1223-16B9 (Hamamatsu)	2.7×1.5 mm	Si p-n diode	UV-NIR	10^{-6}-10^4 lx @ 10^{-14}-10^{-4} A	0.5 μs	Medium
S1721 (Hamamatsu)	TO-5	Si PIN	Vis-NIR	10^{-10}-10^{-4} W @ 10^{-14}-10^{-3} A	50 MHz	High
G1125-02 (Hamamatsu)	TO-5	Schottky	UV-Vis	10^{-13}-10^{-2} W @ 10^{-14}-10^{-3} A	3.5 μs	Medium
S2384 (Hamamatsu)	TO-5	APD	Vis-NIR	Gain=100	1,200 MHz	Medium
IPL10530DAL	TO-5	PD IC	Vis	86 mV/μW/cm^2	65 kHz	Medium
TDET801W (III-V Semi.)	TO-18	Photo-transistor	Vis	0.5 mA @ 5 mW/cm^2	1 μs	Low
L1481 (QT)	TO-18	Photo-darlington	Vis	3 mA @ 0.2 mW/cm^2	75 μs,50 μs	Low

varies perhaps from 10^{12} to 10^9 Ω with temperatures of -10 to +60°C. Consequently, the precise measurement of low light levels needs temperature stabilisation and Peltier devices are available.

6.4 Infra-red Radiation Microsensors

6.4.1 Characterisation of IR detectors

Infra-red radiation extends from a wavelength of 0.75 μm to about 1,000 μm, and lies between the visible light and microwave regions. Infra-red radiation is emitted by all substances with a temperature above absolute zero. The amount of energy in unit time from unit area of a surface at temperature T at a wavelength is given by Planck's radiation law. The wavelength λ_{peak} of maximum output is related to the absolute temperature by

$$\lambda_{peak} \approx \frac{2898}{T} \text{ (in μm)} \tag{6.19}$$

Thus, at a room temperature of 25°C the peak wavelength is 9.7 μm which is in the infra-red region of the electromagnetic spectrum.

There are two types of infra-red detectors: the thermal type and the quantum type. The thermal type includes contacting temperature sensors such as the thermocouple and thermopile (§5.2) as well as the non-contacting pyroelectric detector. In contrast the quantum type has a strong wavelength dependence with faster responses and includes the photoconductive and photovoltaic devices discussed in the last section.

Table 6.6 summarises the types of IR sensors with the photosensitive materials, spectral range, sensitivity and operating temperature.

Table 6.6 Characteristics of some common IR sensors.

Type	Principle	Material	Spectral range (μm)	Detectivity D^* (cm Hz^{½}/W)	Operating temp. (K)
Quantum	Photoconductive[3]	PbS	1 to 3	1×10^9	300
		PbSe	1 to 4.5	1×10^8	300
		HgCdTe	2 to 12	2×10^{10}	77
Quantum	Photovoltaic[4]	Ge	0.6 to 0.9	1×10^{11}	300
		InAs	1 to 3	1×10^{10}	77
		InSb	2 to 5.5	2×10^{10}	77
Thermal	Pyroelectric[5]	TGS	-	2×10^8	300
		LiTaO$_3$	-	2×10^8	300

[3] $D^*(500,600,1)$.
[4] $D^* (500,1200,1)$.
[5] $D^* (\lambda,10,1)$.

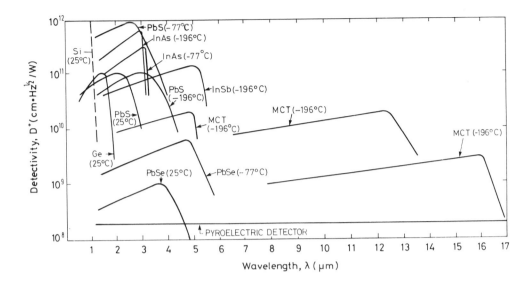

Figure 6.17 Detectivity of infra-red radiation sensors.

The detectivity or detection capacity, D^*, is a measure of the performance of the IR sensors. D^* is related to the effective sensitive area A_{eff} and noise equivalent power (NEP) by

$$D^* = \frac{\sqrt{A_{eff}}}{NEP} = \frac{\sqrt{A_{eff}} \cdot \text{Radiant sensitivity (A/W)}}{\text{Noise current (A/Hz}^{1/2})} \qquad (6.20)$$

Figure 6.17 shows the typical spectral response of IR sensors against wavelength of radiation. Clearly, silicon is an unsuitable material while HgCdTe (MCT) covers a wide range and is practically independent of wavelength.

IR radiation sensors are widely used in industry, agriculture, medicine, physics, chemistry, optical communication and long range remote sensing. The characteristics of each type of device will now be discussed and some applications given.

6.4.2 Photoconductive IR sensors

Thin films of PbS, PbSe and MCT are used to make photoconductive cells in which the incident radiation lowers the resistance of the device. The PbS cells have a peak response at 2.2 μm while PbSe cells have a peak response at 3.8 μm (both at 25°C). MCT cells have a band-gap that may be varied through the composition ratio of HgTe to CdTe and so several spectral response characteristics are available (e.g. 12.2 μm or 16 μm).

Changes in temperature lead to changes in the cell's spectral response, dark resistance and time constant. The dark resistance falls by about 3%/°C and the time constant falls by about 5%/°C. Consequently, it is preferable to run the devices at a constant temperature, e.g. PbS and PbSe at 300 K and MCT at 77 K.

Figure 6.18 shows the basic operating circuit for a photoconductive IR sensor. Typical values of the components are indicated. The output voltage V_{out} is given by,

$$V_{out} = i_s R_d \frac{R_f}{R_g}$$

(6.21)

where R_d is the dark resistance of the device. The response is quite linear with irradiance over the range of 10^{-6} to 10^{-3} W/cm^2.

Photoconductive IR sensors can be used in a variety of applications as shown below in Table 6.7. Some applications, such as gas sensing, are described in other chapters (e.g. §9.1).

Table 6.7 Applications of some common photoconductive IR sensors.

Application	Material:	PbS	PbSe	MCT
Hot metal detector		Yes	No	No
Flame monitor		Yes	No	No
Water content analyser		Yes	No	Yes
FTIR spectroscopy		Yes	Yes	Yes
Film thickness monitor		Yes	Yes	Yes
Fire detector		No	Yes	No
Gas sensing		No	Yes	Yes
IR imaging device		No	No	Yes
Long range remote sensing		No	No	Yes

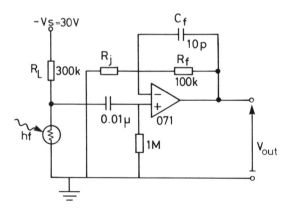

Figure 6.18 Basic operating circuit of a photoconductive microsensor.

6.4.3 Photovoltaic IR sensors

A family of materials is used in photovoltaic sensors to detect the IR (including NIR) spectrum. The common semiconductors that are used are Ge, InGaAs, InAs and InSb with their spectral bandwidths shown in Figure 6.6.

IR photodiodes operate in the same manner as has already been described for UV-NIR p-n photodiodes. The open-circuit voltage V_{oc} varies logarithmically with incident illuminance while the short-circuit current I_{sc} varies linearly. The voltage and current can be measured using the circuits given in Figure 6.14 although a feedback capacitor (~10 pF) should be added to the op-amp circuit to provide noise filtering and thus obviate any dynamic instability.

Typical characteristics of an InGaAs IR radiation photodiode are a radiant sensitivity of 900 mA/W at 1.3 μm, a short-circuit current of 50 nA from an 80 μm diameter window, a dark current of 100 pA at V_R of 5 V and 25°C, a detectivity of 5×10^{12} cm.Hz$^{\frac{1}{2}}$/W at λ_p, device capacitance of 1 to 10 pF at 1 MHz, a fast response time of ~1 ns for a 50 Ω load resistance at V_R of 5 V, and a cut-off frequency of up to 2 GHz. The output current can be increased by increasing the size of the effective sensitive area from 0.005 mm^2 up to 0.8 mm^2 or more. However, there may be a corresponding reduction in the performance with respect to the RC time constant of the read-out circuit.

Table 6.8 illustrates some of the practical applications of these types of IR sensors.

Table 6.8 Applications of some common photovoltaic IR detectors.

Material: Application	Ge	InGaAs	InAs	InSb
Radiation thermometer	Yes	No	Yes	Yes
Hot metal detector	Yes	No	No	No
Gas sensing	No	No	No	Yes
FTIR spectroscopy	Yes[6]	No	Yes	Yes
Laser monitor	Yes	Yes	Yes	No
Photometers	Yes	No	No	No
Laser diode testing	Yes	No	No	No
IR imaging devices	No	No	No	Yes
Long range remote serving	No	No	No	Yes

6.4.4 Pyroelectric sensors

Pyroelectric sensors are a thermal rather than quantum type of infra-red sensor (see Table 6.6). A pyroelectric sensor usually consists of a single crystal of LiTaO$_2$ which is in a polarised state due to an electric field. The crystal surface is always

---

[6] Avalanche type rather than p-n type of Ge photodiode.

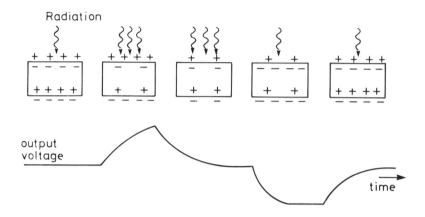

Figure 6.19 Principle of a pyroelectric radiation microsensor.

charged with ions in the air balancing the crystal charge. When the incident radiation level is modified (e.g. increased), the crystal absorbs more radiation and its temperature rises. The increase in temperature reduces the polarisation of the crystal, thus its voltage will increase until the surface charge is once again balanced. A reverse signal is observed when the increase in radiation is removed with the entire process illustrated in Figure 6.19.

LiTaO$_3$ has a very high impedance and so an FET is integrated into the device to buffer the impedance. The response Δy of a pyroelectric detector can be related to its thermal and electrical properties by

$$\Delta y = \frac{\eta \Gamma AR}{\kappa} \cdot \frac{\omega}{\sqrt{(1+\omega^2\tau_T^2)(1+\omega^2\tau_E^2)}} \qquad (6.22)$$

where η is the quantum efficiency, Γ is a pyroelectric factor, A is the sensitive area, R the resistance of the device, ω the angular frequency, τ_T and τ_E the thermal and electrical time constants and κ is the thermal conductivity.

Figure 6.20 shows a basic operating circuit for a pyroelectric detector. The bandwidth is now determined by the circuit time constants with a low frequency cut-off of $1/(2\pi R_1 C_1)$ and a high frequency cut-off of $1/(2\pi R_2 C_2)$.

Pyroelectric sensors have a good temperature stability with a temperature coefficient of output of about 0.2%/°C at 1 Hz. As the sensor output equilibrates to zero, long term fluctuations in ambient temperature are effectively removed.

Pyroelectric sensors can be used as radiation thermometers and gas analysers but a major application is in human body detection (e.g. burglar alarms) and fire detectors. By placing a Fresnel lens in front of the detector, a person walking

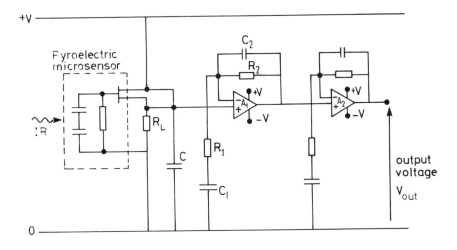

Figure 6.20 Basic operating circuit of a pyroelectric microsensor.

across the field of view will generate a chopped IR signal. An appropriate choice of band-pass filter and comparator will produce a reliable detection system at low cost.

6.4.5 *Relative performance of commercial devices*

Three main types of IR detectors have been discussed, photoconductive, photovoltaic and pyroelectric. There are a large number of commercially available devices for differing application. Table 6.9 lists some commercial devices and gives their characteristic features. The cost is also indicated and those operating at liquid nitrogen temperature are clearly more inconvenient to use.

Suggested further reading

Readers are referred to the following books for additional information on the topic of radiation:

Burcham WE: *Nuclear Physics: An Introduction* (1973). Published by Longman Group Ltd, London. ISBN 0-582-44110-2. (686 pages. Background details on nuclear radiation)

Pankore JI: *Optical Processes in Semiconductors* (1971). Published by Dover Books, USA. ISBN 0-486-60275-3. (422 pages. Good overview of optical properties of semiconductor materials)

Table 6.9 Relative performance of some commercial IR detectors.

Device/ Manufacturer	Package	Type	Range (μm)	Typical output	Response time	Cost
P394A (Hamamatsu)	TO-5	PbS PC (uncooled)	1-2.5	5×10^4 V/W @ λ_{peak}	100-400 μs	Medium
P791 (Hamamatsu)	TO-5	PbSe PC (uncooled)	2-5	8×10^2 V/W @ λ_{peak}	2-5 μs	Medium
P3981 (Hamamatsu)	TO-8	MCT PC (cooled)	2-5	$D^*=(500,1200,1)$ 2.5×10^9	10 μs	High
B2538-01 (Hamamatsu)	TO-8	Ge PD (cooled)	1-1.8	800 mA/W @ λ_{peak} $I_{sc}=1.4\mu$A at 100 lx	1 μs	High
G3476-01 (Hamamatsu)	TO-18	InGaAs PD (uncooled)	0.8-1.7	900 mA/W @ 1.3μm	0.3 ns	High
DP-2101-101 (Sentel)	TO-5	Pyro	7-14	1800 V/W	~100 ms	Low

References

6.1   Burcham WE (1973) *Nuclear Physics*, Longman Group Ltd, London, UK, 686 pp.

6.2   Malm HL, Raudorf TW, Martini M and Zanio KR (1973) Gamma-ray efficiency comparisons for Si(Li), Ge, CdTe and HgI$_2$ detectors, *IEEE Trans. Nucl. Sci.* **20**, 500-509.

6.3   Hamamatsu (Japan) 1990-91 Catalogue on CdS Photoconductive cells.

6.4   Pankore JI (1971) *Optical Processes in Semiconductors*, Dover Publications Inc., New York, USA, 424 pp.

6.5   Sze SM (1985) *Semiconductor Devices, Physics and Technology*, John Wiley & Sons Inc., New York, USA, 523 pp.

Problems

6.1   State the type of sensor that you would employ to measure: (a) fast electrons, (b) neutrons and (c) alpha particles. Give a brief explanation of your choice of sensor in each of the three cases.

6.2   An engineer needs to design a safety light which turns itself on when the ambient light level falls below 5 lux. Suggest a possible radiation sensor to use, together with a possible electronic driving circuit. What parameters does the response time and power consumption of your safety light depend upon? Estimate typical values of the response time and power consumption.

6.3 What kind of radiation sensor would you use in the following applications: (a) to detect the position of a hot rotating metal plate used to seal plastic containers as they pass by on a conveyor belt in a manufacturing process; and (b) to detect a stationary object emitting a very low light signal with an intensity below the threshold of the human eye, such as photoluminescent light generated by bacteria in a petri dish?

6.4 Discuss the general advantages and disadvantages of using non-contacting radiation microsensors rather than contacting microsensors. Illustrate your answer with two examples of specific applications of radiation microsensors in a manufacturing process. *(Please do not use the examples given in earlier questions).*

7. Mechanical Microsensors

Objectives

- [ ] To introduce the topic of mechanical microsensors

- [ ] To describe the basic types of mechanical microsensors

- [ ] To examine the underlying principles

7.1 Mechanical Measurands

7.1.1 Basic considerations

Mechanical sensors form perhaps the largest class of sensor because of their widespread applicability. There are a large number of mechanical measurands to consider, from acceleration to yaw. Table 7.1 lists over 50 measurands that would normally be classified as mechanical in nature [7.1]. The list is in alphabetical order but is by no means complete. In fact, many of the measurands listed here could be subdivided further, for example, friction into static or dynamic friction.

Clearly, there is a need to reduce this large number of measurands into a smaller, more manageable scheme that could be used to define the genus of mechanical sensors. Table 7.2 shows a smaller number of mechanical measurands which are used here to define the most important classes of mechanical sensors. This list is based upon a scheme proposed elsewhere [7.2]. An advantage of this abbreviated list is that it covers the main types of mechanical measurands which thus includes many of the specialised terms (e.g. displacement covers various measurands such as deflection or movement). These twelve categories define the most common types of mechanical sensors.

The Japanese market for solid-state sensors was examined in Chapter 1. The number of mechanical units sold and their market value is given in Table 7.3. As we can see, the Japanese market for mechanical (solid-state) sensors in 1987 was about £242 million with over 246 million units sold. Clearly, the world market for mechanical sensors in 1993 is many times larger than that for Japan in 1987, and is probably well in excess of £1 billion.

Table 7.1 Mechanical measurands. Adapted from [7.1].

Acceleration, angular	Flow, gas	Momentum, linear	Stiffness
Acceleration, linear	Flow, liquid	Movement	Strain
Acoustic energy	Flow-rate	Orientation	Stress
Altitude	Force, simple	Path length	Tension
Angle	Force, complex	Pitch	Thickness
Attraction	Frequency	Position	Torque
Compliance	Friction	Pressure	Touch
Contraction	Hardness	Proximity	Velocity, angular
Deflection	Immersion	Roll	Velocity, linear
Deformation	Inclination	Rotation	Vibration
Density	Length	Roughness	Viscosity
Diameter	Level	SAW	Volume
Dipole alignment	Mass	Shape	Wavelength
Displacement	Microbend	Shock	Yaw
Elastic properties	Momentum, angular	Sound level	

Table 7.2 Classification of main mechanical measurands. Adapted from [7.2].

1. Position, displacement	7. Stiffness, compliance
2. Velocity, speed	8. Mass, density
3. Acceleration	9. Flow-rate
4. Force, torque	10. Shape, roughness
5. Stress, pressure	11. Viscosity
6. Strain	12. Other (Acoustic/Ultrasonic)

Table 7.3 Japanese market for some types of mechanical (solid-state) sensors in 1987.

Sensor type	Units sold (million)	Value (£ million)
Displacement	189.0	133
Level[1]	2.5	7
Speed (rotation)	0.4	5
Vibration	2.2	8
Force or weight	0.2	3
Pressure	3.0	27
Sound	49.0	11
Total:	246.3	194

[1] A level sensor may be regarded as a displacement sensor.

7.1.2 Principles of mechanical microsensors

Mechanical parameters, such as position, velocity, and force, can be measured using familiar large sensors. However, the scope of this book is restricted to those sensors that convert a mechanical signal into an electronic one using solid-state microtechnology. This definition still covers some of the traditional principles to measure the extrinsic properties of microstructures, such as the capacitance or inductance, see Table 7.4. In addition, it covers the variation of material (intrinsic) properties employed in measurement.

Table 7.4 Properties employed in some typical mechanical microsensors.

Nature	Property/principle	Examples
Intrinsic:	Resonant (acoustic)	Micromass gauge
	Resonant (elastic)	Microflexural systems
	Resistive	Strain gauges
	Piezoresistive	Pressure gauges
	Piezoelectric	Pressure gauges
	Dielectric	Strain gauges
Extrinsic:	Capacitive	Pressure gauges
	Inductive	LVDTs
	Reluctive	Hall position sensors
	Magnetic coupling	Resolvers
	Optical coupling	Optical encoders

In the last two chapters we have discussed the use of the properties of various (crystalline) materials to measure thermal or radiation parameters. The relations between the physical properties of crystalline materials are illustrated in Figure 7.1 [7.3]. Figure 7.1 shows the primary relations between stress, temperature and electric field in a crystalline material. The names of the associated effects are written alongside the connecting lines. The route taken from, for example, the stress node, must end at the electrical node to have a sensor rather than an actuator. However, the route may be direct or indirect. For example in an indirect route, temperature could be converted to strain by thermal expansion which is then converted to an electrical signal via a piezoelectric effect. The figure thus illustrates the important effects that can be used in mechanical microsensors which exploit crystalline materials. Some of these effects are also found in non-crystalline materials, e.g. polymers.

7.1.3 Classification of mechanical sensors

Mechanical microsensors will be described here in terms of the twelve classes listed in Table 7.2. A sensor may measure an absolute quantity or a differential quantity (e.g. position or displacement). A sensor may measure linear or angular parameters, so a position sensor can measure either a linear displacement or

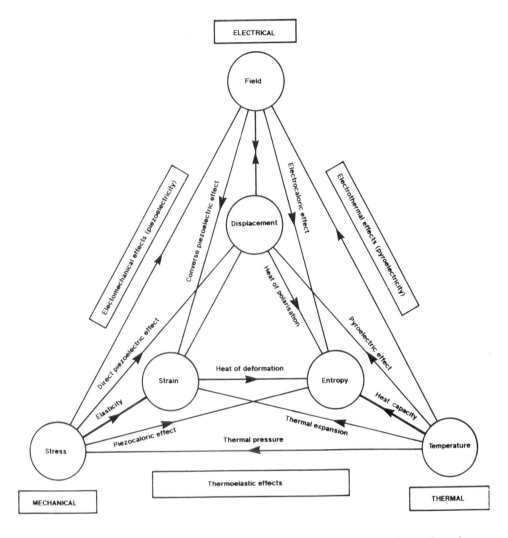

Figure 7.1 Some physical relationships between mechanical, thermal and electrical properties of crystalline materials. Redrawn from [7.3] by permission of Oxford University Press, UK.

angular rotation. The type of microsensor can then be described more precisely by its principle of operation, such as piezoresistivity. For example, pressure may be measured by a piezoresistive pressure sensor. Figure 7.2 has been constructed using this classification scheme.

One particular material, above all others, is arguably the most important in the field of micromechanics and that is silicon. Like steel and concrete in macrostructures, silicon is a common choice of building material in microstructures. The main reasons are that silicon is already widely used in microelectronics, its properties (as evident from Chapters 4 and 5) are well known and full integration is possible. Consequently, some of the mechanical properties

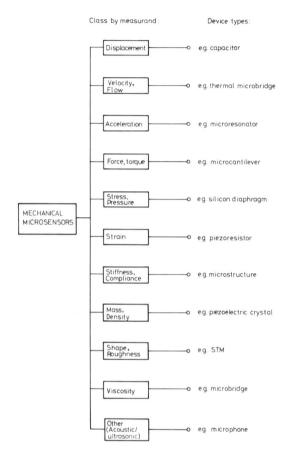

Figure 7.2 Classification scheme of mechanical microsensors.

of silicon that make it suitable for microstructures are discussed in the next section.

7.2 Mechanical Properties of Silicon Microstructures

7.2.1 Mechanical structures

Engineers have designed, constructed and operated mechanical structures for centuries. Consequently, there are well-known design codes for constructing many mechanical structures from large buildings and bridges to small enclosures. There are a variety of mechanical parameters which are important to the design of the desired structure. If we consider the most basic of structures, such as beams, bridges, diaphragms and flexures, then we can rapidly identify some of the key mechanical parameters. The most obvious design parameters are the physical dimensions of a structure, e.g. its width, height or depth, thickness or radius for axisymmetric structures. In fact, we often identify dimensionless ratios of these parameters in order to simplify the design rules. However, this is strictly valid only if some kind of simple scaling theory is obeyed.

Secondly, the designer often has some choice of possible material to use with its desirable material properties, e.g. a high elastic modulus and low density. However, the choice may be severely limited due to some processing considerations or simply the cost of the material. As silica (sand) and iron (iron pyrites) are abundant in nature, we can use them to make mechanical structures at reasonably low unit costs.

Thirdly, there are a set of mechanical parameters that can be calculated but are essentially defined by our choice of physical dimensions and the material properties. Examples of these are the structure's mass, spring constant, damping coefficient, strain, natural frequency, and moment of inertia.

Finally, the designer must consider the applied inherent loads upon the structure, such as the applied force, torque, stress, pressure and the response of the structure to them. In statics, the structure is stationary and so the response is independent of time, e.g. the deflection of a beam under constant load, whereas in dynamics, the response of the structure varies with time and hence the harmonic content of the load and transfer function of the structure. These parameters and properties are summarised in Table 7.5 below.

Table 7.5 Key mechanical parameters and properties in design.

Type	Examples
Physical dimensions	Width (w), height (h), thickness (d), radius (a).
Material properties	Elastic modulus (E_m), Yield strength (Y), Poisson's ratio (v_m), Density (ρ_m), Viscosity (η_m).
Calculable parameters	Mass (m), Spring constant (k_m), Damping coefficient (b_m), strain (ε_m), Natural frequency (ω_0), Moment of inertia (I).
Load parameters	Applied force (F), Applied torque (τ_m), Stress (σ_m), Pressure (P), Impulse (I_m).
Response parameters	Lateral deflection (y), Angular deflection (θ), Sensitivity (S), Resonance (Q), Band-width ($\Delta\omega$).

The theoretical and practical relationships between the applied load and the resultant response are well known for many mechanical structures. Indeed, this information is contained in standard text-books or design codes for different materials. In many cases, a linear elastic theory may be applied. For example, in a one-dimensional structure this is simply Hooke's law where the displacement Δx of an ideal material is directly proportional to the applied force F,

$$F = k_m \Delta x \qquad (7.1)$$

For distributed loads, the applied force becomes a stress σ_m and the response becomes the strain ε_m, where

$$\sigma_m = E_m \varepsilon_m \qquad (7.2)$$

When mechanical structures obey a linear theory, the spring constant k_m and Young's modulus E_m are independent of the applied load. Consequently, the deflection y of a cantilever beam of length l to a simple force applied at its end can be written as,

$$y = \left[l^3 / 3 E_m I\right] F \quad \text{or} \quad y \propto F \tag{7.3}$$

The gain of this system (force/deflection) is hence now a physical stiffness constant ($l^3/3EI_m$) which only depends upon the geometrical size and material properties of the beam.

7.2.2 Micromechanical scaling

Using micromachining techniques, it is possible to fabricate mechanical structures that have a dimension on the micron rather than millimetre or metre scale. The question then arises as to whether the fundamental laws of mechanics are obeyed when the scale of the structure decreases from the macro to the micro. In other words, will the deflection of a cantilever beam only 1 μm thick satisfy the linear elastic theory described by equation (7.3)?

The effect of geometric scaling on a parameter ultimately depends upon the underlying physical principle. However, it is useful to examine the dimensions of parameters (or their ratios) to determine the ideal rules of isotropic scaling. Some parameters, such as flow-rate, surface tension and specific heat capacity, are ideally independent of the scale, while others, such as density, contain size-dependent terms and therefore they do not scale.

Dimensional analysis can be used to determine the geometrical scaling factor of quantities or parameters, e.g. displacement. Thus, all lengths scale by a factor K and we can calculate the scale factor for other mechanical parameters. For example, scaling at constant stress produces the relationships for a cantilever beam as shown in Table 7.6.

Table 7.6 Geometric scaling factors of a cantilever beam (constant stress).

Parameter	Symbol	Factor
Beam force	F	K^2
Spring constant	k_m	K
Deflection	y	K
Beam stress	σ_m	1
Mass	m	K^3
Natural frequency	ω_o	K^{-1}

Consequently, a reduction in the size of a cantilever structure will theoretically increase its natural frequency, but reduce its spring constant, deflection and mass. The considerable reduction in mass ($\propto K^3$) means that the materials cost is lower

Table 7.7 Dimensions of physical sensing parameters.

Physical parameter	Electromagnetics	Electrostatics
Capacitance	$L^{-1}T^2\mu^{B-1}$	$L\,\varepsilon$
Inductance	$L\,\mu$	$L^{-1}T^2\,\varepsilon$
Resistance	$L\,T^{-1}\mu^B$	$L^{-1}T^{-1}$
Electric current	$M^{1/2}L^{1/2}T^{-1}\mu^{B-1/2}$	$M^{1/2}L^{3/2}T^{-2}\,\varepsilon^{1/2}$
Potential difference	$M^{1/2}L^{3/2}T^{-2}\mu^{B-1/2}$	$M^{1/2}L^{1/2}T^{-1}\,\varepsilon^{1/2}$

and, perhaps more importantly, the response time of the mechanical structure is greatly reduced by its higher natural frequency.

Geometrical scaling may also affect the relative performance of various sensing principles. Table 7.7 shows the dimensions of various physical principles in terms of mass (M), length (L) and time (T). The scaling effect is such that, for example, electrostatic methods are preferable to electromagnetic ones at the micron level.

Finally, dissipative or frictional phenomena also have an associated geometric scaling factor. The common parameters are:

- viscous damping (K^2);
- mass coulomb damping (K^3);
- elastic coulomb damping (K^2);
- surface adhesion (K^2).

In practice, microstructures do not necessarily scale as expected, i.e. isotropically. For example, viscous damping in a small cavity is lower than expected when the mean free path of air molecules approaches the size of the cavity (the Knudsen effect). Thus, an isotropic linear theory can break down on the microscopic scale. Ideally, microstructures should be designed so that the linear theories of mechanics still apply. This is usually the case for the silicon micromechanical structures described in this chapter, and leads to straightforward mechanical models.

7.2.3 Silicon micromechanical structures

Our microelectronics industry today is primarily based upon utilising the excellent electronic properties of single-crystal silicon. As described in Chapters 3 and 4, silicon can be processed conveniently to make microelectronic devices or with added steps, to make microsensors. Yet, silicon is a versatile material as it also has some excellent mechanical properties which can be used to fabricate precise and reliable mechanical microstructures [7.4].

Table 7.8 illustrates some of the mechanical properties of single-crystal silicon together with those for more conventional common materials (see §4.5). From Table 7.8, it is apparent that single-crystal silicon is similar in Young's modulus to iron and steel. Although silicon is a brittle material and cannot deform plastically like some metals, this produces a reliable and precise material as it either has near-perfect mechanical integrity or it fails catastrophically! In practice, the fragility of

Table 7.8 Mechanical properties of single-crystal (SC) silicon and some other materials. Adapted from [7.4], © 1982 IEEE.

Property: Material	Yield strength (GPa)	Knoop hardness (10^9 kg/m^2)	Young's modulus (GPa)
Si (SC)	7.0	0.8	190
SiO$_2$ (fibres)	8.4	0.8	73
Si$_3$N$_4$ (SC)	14.0	3.5	385
Iron (whisker)	12.6	0.4	196
Iron (wrought)	0.15	-	197
Steel (max. strength)	4.2	1.5	210
Diamond (SC)	53.0	7.0	1,035

silicon is determined by its surface roughness and defect density. This phenomenon is shown in Figure 7.3 [7.5]. Careful polishing of a silicon wafer produces a higher fatigue strength, and this can be done mechanically, chemically (§3.3) or through a combination of both. The fatigue strength of silicon can be further enhanced by placing it under an external compressive load or using Si$_3$N$_4$ films which impart a residual compressive stress.

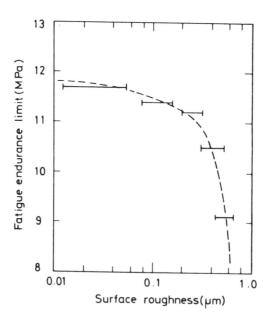

Figure 7.3 Effect of surface roughness on the fatigue endurance of single-crystal silicon [7.4] (© 1982 IEEE).

Silicon micromachining techniques can be used to fabricate a variety of mechanical structures. The bulk micromachining of single-crystal silicon can produce diaphragms, bridges, beams etc. (§3.3). In practice, the conventional theory of elasticity can often be applied to describe the mechanical behaviour of these structures. In addition geometric scaling factors can be used but there are several facts that need to be considered when making microstructures:

- accurate dimensional tolerances are essential for a good manufacturing process. This may be difficult with some types of etching;
- the use of dopants or other thin film materials may lead to considerable residual stresses that can cause the buckling of a structure;
- fatigue endurance will depend upon the surface roughness and hence the quality of the etching process;
- viscous damping depends non-linearly upon scale when the Knudsen region is entered.

Polycrystalline silicon is often used in the surface micromachining of structures. The mechanical properties of polysilicon are predominantly determined by the grain sizes. In these microstructures, the following rule should be observed to retain mechanical integrity:

- the smallest physical dimension of the structure should be much greater than the typical grain size.

Table 7.9 shows the average grain size in LPCVD polycrystalline silicon at various deposition temperatures. An average grain size of below 100 nm permits the use of a micro-sized microstructural dimension while obeying simple elastic theory.

Table 7.9 Average grain sizes of LPCVD polycrystalline silicon.

Temperature (°C)	600	625	650	675	700	725
Grain size (nm)	55	87	72	74	73	86

Microflexural structures have many applications in the field of sensing and can be fabricated using bulk (§3.3) and surface (§3.4) micromachining [7.6]. The simplest of microflexural structures is the cantilever beam described earlier. The addition of a capacitative plate allows the beam to be electrostatically driven (as found to be beneficial on dimensional grounds). Figure 7.4(a) shows a schematic diagram of a vertical microresonator that could be made using bulk micromachining, while 7.4(b) shows a lateral microresonator that requires surface micromachining. These microresonator systems can be modelled as a second order system with a characteristic mass m, spring-constant k_m, damping coefficient b_m, load (force) function F and response y:

$$m\ddot{y} + b_m\dot{y} + k_m y = F(t) \qquad\qquad (7.4)$$

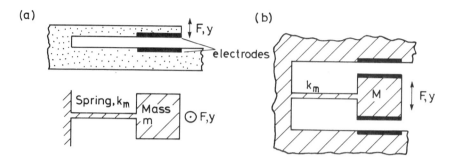

Figure 7.4 Basic design of a microflexural structure: (a) vertical, and (b) lateral.

The mechanical driving force F(*t*) is in this case electrostatic. Thus, the larger surface area of the vertical resonator often makes it more desirable than the lateral resonator. In addition, the asymmetric design of a simple cantilever beam can produce an undesirable out of plane motion. Structural compensation is commonly used to produce deflection-independent spring-constants, for example, a folded flexure or crab-leg design as shown in Figure 7.5. The spring constant k_m of a Hammock microflexural structure depends more strongly upon the deflection y than a simple cantilever beam. In fact, the relationship between the spring constant k_m and the deflection y is a quadratic polynomial. This so-called hard spring causes a non-linear system which is described by Duffing's equation. In contrast, the folded and crab-leg flexures are linear and tend to have smaller maximum stresses (e.g. 200 to 800 MPa) for the same deflection (e.g. 2 to 10 μm). The characteristics of these designs are summarised in Table 7.10 below.

Some advantages of using microflexural structures are that there is no mechanical contact (hence no traditional wear mechanism), damping is reduced in the Knudsen region, substantial displacements can be achieved, characteristics are predicted by conventional mechanical theory, high resonant frequencies (~50 kHz) and so large band-widths, and good in-plane stability. Two main problems in using microstructures are the sensitivity to residual stresses and low capacitance for capacitive drive systems. The former is overcome through careful preparation while the latter may necessitate the use of integrated FET pre-amplifiers.

Table 7.10 Mechanical characteristics of microflexures.

Parameter	Hammock	Folded	Crab-leg
Bending deflection, y	Axial and bending stress	Bending stress	Bending stress
Spring constant, k_y	Non-linear dependence on y	Constant and independent of y	Constant and independent of y
Axial deflection, x	Axial stress only	Axial and bending stress	Axial and bending stress
Spring constant, k_x	Stiff: $4EA/L$	Quite stiff	Stiff

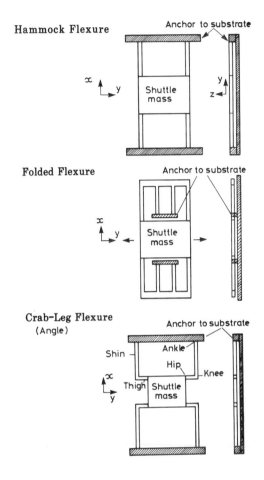

Figure 7.5 Simple, folded and crab-leg microflexural designs.

The ability to use silicon as a mechanical structure has led to the development of many new types of microsensors. These silicon micromechanical sensors can be manufactured in high volume at low cost using traditional silicon processing techniques. Thus, silicon mechanical sensors have rapidly established a large market in a wide range of fields, including automotive and aerospace.

7.3 Displacement Microsensors

7.3.1 Basic considerations

A displacement sensor is defined here as a sensor which can measure a displacement (i.e. a change in position), detect motion, or measure the position of an object. A proximity sensor measures the spatial closeness of two objects and is classified here as a type of displacement sensor.

Displacement sensors may also be classified into contacting and non-contacting by the nature of their sensing principle. Contacting sensing methods may require, for example, the mechanical fastening of a sensing shaft to the point of measurement. The position (e.g. linear or angular) of the sensing shaft is then transduced into an electrical signal. Conventional contacting displacement sensors commonly rely on resistive (e.g. a linear potentiometer), capacitive (e.g. piezoelectric), inductive (e.g. LVDTs, resolvers, synchronisers and encoders), magnetic (e.g. Hall effect and encoders) and optical principles (e.g. encoders).

In contrast, non-contacting sensing methods detect the displacement of the object itself. Conventional non-contacting displacement sensors commonly rely on capacitive, inductive or magnetic techniques, or ranging using lasers, infra-red beams, microwave beams and ultrasonic waves. A good description of the various types of conventional mechanical sensors can be found in reference [7.7].

Displacement sensors are used in a wide variety of applications and so are fundamental to general engineering practice in which the dimensions or tolerances must be established and maintained. The size of the displacement varies enormously from application to application. Subnanometric displacements may require the use of an X-ray interferometer [7.8] at one end, micrometric displacements may require the use of a laser interferometer, millimetric displacements are sensed by many common types of displacement sensors, metric displacements may require a short-wavelength microwave device and over kilometres one may use a longer one (radar).

Displacement transducers are used in many everyday products and, in fact, a switch may be regarded as a simple kind of position sensor. Displacement sensors are employed in traffic lights, burglar alarms, cars (e.g. switch for interior light), automotive systems, robots and medicine (e.g. monitor for foetus in pregnant women). Most of these examples show the use of a displacement sensor as a monitor. However, another large application area is that in open or closed loop control. For example, optical encoders are commonly used to measure the position of an actuator or drive shaft. Consequently, actuators and drive systems can be servo-controlled by position (and velocity) feedback. This is particularly important in manufacturing industry where automative or robotic processes are essential.

7.3.2 *Capacitive and inductive displacement sensors*

Capacitive and inductive proximity sensors are widely available but are essentially miniature rather than microsensors. These sensors measure the presence of an object which modifies its output capacitance or output inductance and operates in the range of about 0.1 to 10 mm. The capacitive measurement of displacement is widely used in microsensors but not usually in order to measure displacement but other related measurands. For example, displacement is measured capacitively in many micromechanical structures, such as cantilevers, diaphragms and resonant microflexures, to measure indirectly the:

- acceleration;
- force/torque;
- pressure/stress.

These types of mechanical microsensors are discussed later on in this chapter.

7.3.3 Optical displacement sensors

A variety of optical proximity sensors are commercially available. The detection principle consists of either (i) interruption of a direct beam, (ii) specular reflection or (iii) diffuse scattering off a surface. Figure 7.6 shows the schematic layout of a diffuse scan opto-switch. The photodiode emits IR radiation which is scattered off the object (~ 1 to 5 mm away) and collected by a phototransistor. For example, Honeywell manufacture a non-focused reflective opto-switch (HOA1397-2) about 4.5 by 4.7 mm square. The measurement of the position rather than the presence of an object is more difficult. The basic reflection method uses a phototransistor pick-up and so produces a non-linear output that depends upon the flatness of the reflecting object. This method is not very accurate and the occulation of a collimated laser beam which is detected by a LED array would be better. However, the use of a laser beam in either a triangulation or interferometric method may hardly be regarded as a microsensor.

Optical and indeed magnetic encoders are commonly used to measure the shaft angles in non-contacting sensors but recently it was reported that a small photoelectric sensor could be used to measure the angle or inclination of a surface. The principle is simple. A bubble trapped inside a small hemispherical spirit level moves about with the angle of inclination. Light from an LED is reflected off it and casts a shadow on four photodiode diodes. The output from the four photodiodes is then related to the tilt angle and direction via a calibration curve or

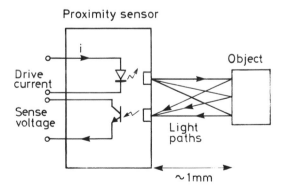

Figure 7.6 A solid-state diffuse scan optical proximity sensor.

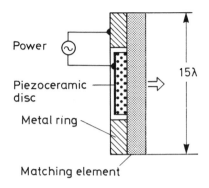

Figure 7.7 An ultrasonic proximity sensor head. Adapted from [7.10].

look-up table [7.9]. Over a range of 0 to 20° tilt for the cardinal point, an error of better than 8% has been achieved.

7.3.4 *Ultrasonic displacement sensors*

Ultrasonic proximity sensors offer several advantages over other types of proximity sensors. First, they are non-contacting and will detect materials which are conducting, non-conducting, ferrous and non-ferrous. Secondly, they transmit high frequency sound waves (typically ~50 kHz in air) and so the signals are less sensitive than optical devices to dirty environments. Finally, the sensing range is much greater at perhaps 5 m. Figure 7.7 shows the general arrangement of an ultrasonic sensor. The piezoelectric element, comprised of lead titanate and zirconate, both generates the outgoing signal and collects the returning echo. Ultrasonic sensors are commonly used to measure proximity [7.10] and distance [7.11]. They can also be used to measure the level of liquids and are commercially available at low cost.

7.3.5 *Pyroelectric and other devices*

Pyroelectric detectors have already been described as IR radiation sensors (§6.4.4). However, they can be modified to measure easily the presence of a person through the addition of a Fresnel lens and suitable electronic circuitry. The compound Fresnel lens both focuses the IR light onto a pyroelectric crystal and modulates the signal intensity as the person moves across the field of view. The differential measurement removes the need to measure the amplitude of the signal and hence avoids problems with drift and temperature dependence.

Low cost pyroelectric proximity sensors are used in, for example, burglar alarms and are commercially available. The use of other material properties to measure displacement is possible, such as piezoresistance and piezoelectricity; however, as we shall see, these contacting methods are used for other types of microsensors (e.g. roughness, strain).

7.3.6 Concluding remarks

The measurement of displacement is clearly of fundamental importance in engineering.   Many solid-state sensors have been developed to measure the proximity or position of objects. Yet, mechanical microsensors more commonly use displacement as an intermediate parameter to measure some other measurand. The principal reason is that using silicon microstructures (or its material properties) to measure displacement results in a very small dynamic range of 0.1 to 10 µm. This restricts their use to specialised microengineering applications such as microprobes (in roughness measurement, see §7.8.2) or feedback sensing mechanisms in microactuators.

7.4 Velocity and Flow Microsensors

7.4.1 Basic considerations

Velocity v is the rate of change of displacement x with time t and is a vector quantity. In contrast speed is the magnitude of velocity and is a scalar quantity. For a linear or translational system, the velocity and speed are,

$$\text{Velocity, } v = \frac{dx}{dt}; \quad \text{Speed, } |v| = \left|\frac{dx}{dt}\right| \tag{7.5}$$

In a rotational system, the angular velocity ω is the rate of change of orientation θ with time and is related to the translational velocity by the radial distance a.

$$\omega = \frac{d\theta}{dt} \quad \text{and} \quad \omega = \frac{v}{a} \tag{7.6}$$

Thus, the velocity of an object can be calculated from the first time derivative of its displacement or alternatively from the integral of its acceleration a,

$$v = \frac{dx}{dt} \quad \text{or} \quad \int_0^t a \, dt'$$

$$\omega = \frac{d\theta}{dt} \quad \text{or} \quad \int_0^\theta \alpha_\theta \, dt' \tag{7.7}$$

Integration of an acceleration term provides a differential measurement and not an absolute one.

The velocity of some objects can also be derived from a pressure-altitude sensor. For example, the rate of descent of an aircraft v_h is related to its change in height h and pressure P by,

$$v_h = \left(\frac{dh}{dP}\right)\left(\frac{dP}{dt}\right) \tag{7.8}$$

Thus, velocity sensing methods include those discussed already under displacement sensing (e.g. capacitive coupling and optical encoders), accelerometers (e.g. resonant microflexures) and altitude pressure sensors (e.g. silicon diaphragms).

Although the motion of a rigid body is described in terms of its velocity, the motion of a fluid (liquid or gas) is described by its flow. The flow-rate of a fluid is the rate of motion of a fluid with time. It is usual to define the mass flow-rate Q_m and volumetric flow-rate Q_v when discussing the fluid mass m or volume V passing by a point in space.

$$Q_m = \frac{dm}{dt}; \quad Q_v = \frac{dV}{dt} \tag{7.9}$$

Both the mass and volumetric flow-rates are related to the velocity of the liquid. The mass flow-rate of a fluid of density ρ_m is related to its volume V by,

$$Q_m = \frac{dm}{dt} = \frac{d(\rho_m V)}{dt} = \rho_m \frac{dV}{dt} + V\frac{d\rho_m}{dt} \tag{7.10}$$

The first term relates to the volumetric flow-rate whereas the second term represent a change in density with time. Thus, for incompressible fluids the relationship simplifies to

$$Q_m = \rho_m \frac{dV}{dt} = \rho_m Q_v = \rho_m A v \tag{7.11}$$

where A is the cross-sectional area of the flow. The velocity v of a fluid can then be calculated from its mass flow-rate or volumetric flow-rate and vice-versa.

There are also a variety of conventional sensing methods used to measure flow; these include differential-pressure flow sensing (e.g. the orifice plate, venturi tube, Pitot tube), mechanical flow sensing (e.g. rotary or spring-loaded vanes, cantilever beams), thermal flow sensing (e.g. Thomas flow meter, hot-wire anemometer), ultrasonic flow sensors (e.g. Doppler shift, or transit time), magnetic flow sensors (e.g. back e.m.f.), and oscillating-fluid flow sensing (e.g. vortex shedding/Karman street). But in this section we will describe some of the novel velocity or flow-meters that have been fabricated using micromachining techniques. Readers interested in conventional velocity or flow sensors can refer to almost any book on sensors or transducers, e.g. [7.7].

7.4.2 Thermal flow microsensors

The first thermal flow-meter was described by C.C. Thomas in 1911 and measured the heat transferred between two points in a moving liquid. The schematic arrangement is shown in Figure 7.8 (a). A heating element is immersed in the fluid and heat is dissipated into the fluid P_h. The temperature of the fluid is measured before and after the heater by thermocouples (T/C). The mass flow-rate can be found from,

$$Q_m = [P_h/c\Delta T] = [P_h/c(T_2 - T_1)]$$ (7.12)

where c is the specific heat capacity of the fluid. This arrangement is somewhat impractical as the heating wire impedes the flow being measured. Instead the heating element can be wound around the wall, Figure 7.8(b).

The precise equation relating the mass flow-rate and the temperature difference is more complicated in practice because it is determined by the heat transfer coefficient across the boundary-layer. A silicon micromachined version of this has been reported by Johnson and Higashi [7.12]. Anisotropic etching was used to form two bridges with thin film resistors, see Figure 7.9(a). There are two identical thin film thermistors on each end of the structure and one heating element split between the two bridges. The advantage of this structure is that the power required for the heater is very low (about 15°C per mW) and it has a very

(a)

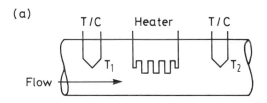

(b)

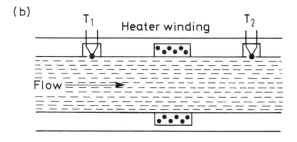

Figure 7.8 Thermal flow sensors: (a) Thomas flow-meter, and (b) boundary-layer flow-meter.

low heat capacity which results in a thermal time constant of only 3 ms. The microsensor can measure flow velocities of air up to 30 m/s in channels of 5 µm by 250 µm and is limited by the onset of turbulent flow. Figure 7.9(b) shows the voltage output from a Wheatstone bridge configuration (followed by a differential input instrumentation amplifier) against mass flow-rate for air, carbon dioxide and methane. The non-linear output can be calibrated using look-up tables or mathematical functions.

The design has been further improved by Stemme [7.13] by placing the heating

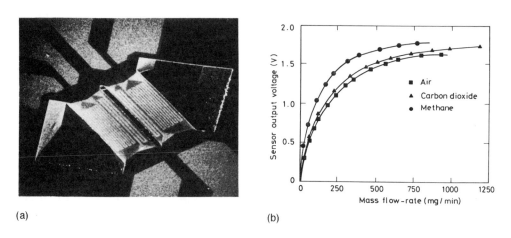

(a)   (b)

Figure 7.9 (a) Micrograph of a silicon microbridge thermal flow sensor and (b) sensor output vs. flow-rate [7.12].

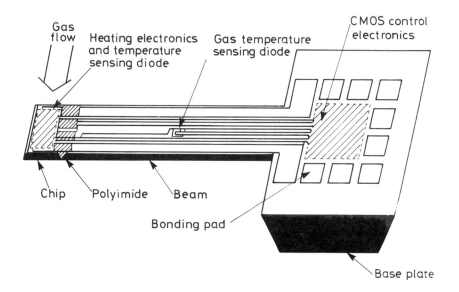

Figure 7.10 A cantilever silicon thermal flow sensor [7.13] (© 1986 IEEE).

resistor on the end of a cantilever beam and thermally isolating it with a polyimide layer. The temperature differential of the gas is measured by two thermodiodes (§5.5) with CMOS control electronics integrated later, Figure 7.10 A thermal time constant of 50 ms was reported with a gas flow velocity range of 0 to 30 m/s. The polyimide layer typically doubled the flow sensitivity by halving the electrical power dissipation to about 50 mW at a flow speed of 10 m/s.

More recently, a thermal flow microsensor has been reported which uses thermocouples (§5.2) rather than thermodiodes and behaves like a weather vane. The device consists of four diffused resistors to heat the sensor, and four 22-element thermally diffused thermopiles with a sensitivity of 13 mV/K. Figure 7.11(a) shows the schematic chip-layout [7.14] which was made using bipolar IC processing. The device was run at a constant temperature differential ($\Delta T = 12$ K) and Figure 7.11(b) shows the thermopile output voltages against flow angle. The power consumption is approximately proportional to the square root of the flow-rate over a range of 0 to 25 m/s.

A similar method has also been used to measure liquid rather than gaseous flow-rates. However, they generally measure lower flow-rates and care is needed to make sure that the liquid does not contaminate the microstructure. Fouling of the surface leads to a change in the heat transfer characteristic and thus the power dissipated.

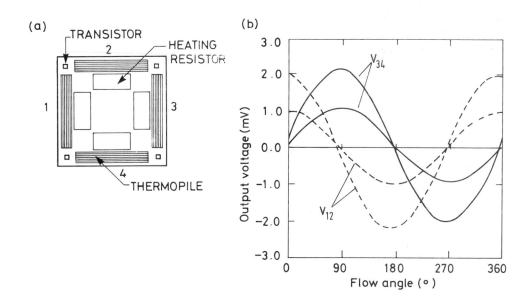

Figure 7.11 A capacitive flow direction microsensor: (a) structure, and (b) voltage output against flow direction for both thermopile pairs [7.14].

Table 7.11 gives the specifications of two commercial silicon microbridge mass airflow sensors made by Honeywell. Notice that the power consumption is only 30 mW, response time 3 ms and hysteresis less than 1%.

Table 7.11 Specifications of commercial (Honeywell) microbridge air flow sensors [7.15].

Model number: Specifications	AWM3100V	AWM3300V
Range (ml/s)	0 to 200	0 to 1,000
Power consumption (mW)	30	30
Output voltage (V DC)	5.00 @ 200 ml/s	5.00 @ 1,000 ml/s
Null voltage (V)	1.00 ± 0.05	1.00 ± 0.10
Repeatability (% FSO)	±0.5	±1.0
Response time (ms)	3.0	3.0
Operating temp. range (°C)	−25 to +85	−25 to +85
Weight (g)	10.8	10.8
Cost (£)	<100	<100

7.4.3 Capacitive flow microsensors

Thermal flow microsensors form the largest class of flow microsensors. Yet there has been recent interest in the use of other transduction principles. Figure 7.12(a) shows the cross-section of a silicon micro-flowmeter [7.16]. The structure was fabricated using bulk silicon micromachining and boron etch-stops (§3.3). Gas flows into the inlet at pressure P_1, passes through a silicon flow channel and leaves with the outlet at pressure P_2. The gas flow-rate Q_m generates the pressure differential $(P_1 - P_2)$ across the fluidic conductance G_f of the channel, where

$$Q_m = G_f(P_1 - P_2) \tag{7.13}$$

The fluidic conductance G_f is determined by the channel dimensions and the viscosity of the gas η_m. For viscous laminar flow of the gas through the channel, the conductance is constant and approximately given by Poiseuille's equation [7.17],

$$G_f = \frac{\pi a^4}{\delta \eta_m l} \tag{7.14}$$

where l is the length of the channel, a the channel radius (approximating the channel to a tube). The differential pressure is measured by a capacitive pressure sensor, Figure 7.12(b). The p^{++} boron-doped silicon membrane forms the deflection plate with a zero flow capacitance of 6.3 pF. The capacitance was measured using an integrated CMOS switched-capacitance circuit which simply

(a)

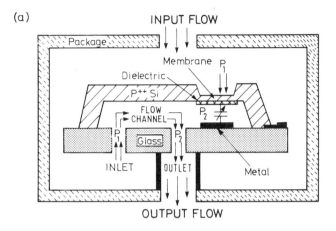

(b)

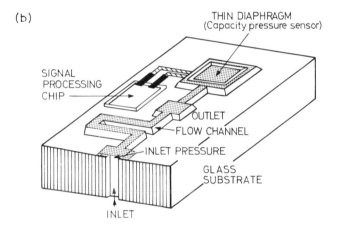

Figure 7.12   A capacitive differential pressure flow microsensor: (a) cross-section, and (b) general layout [7.16].

integrates the charge difference on the sensing capacitance C_s and a reference capacitance C_{ref},

$$V_{out} = \frac{(C_s - C_{ref})}{C_f} V_{ref} \tag{7.15}$$

where C_f is the feedback capacitance in an op-amp circuit and V_{ref} is the voltage reference pulse height. A resolution of 1 fF was reported which corresponds to a change in pressure of 0.13 Pa. The sensor has some self-test capability as the pulse-width and input voltage can be systematically varied to check the pressure sensitivity and transfer function, respectively. Figure 7.13 shows the characteristic performance, in this case linear, of the capacitance flow microsensor at small

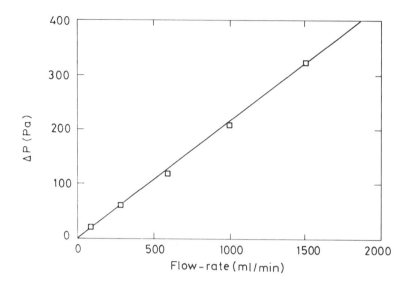

Figure 7.13 Effect of flow-rate on the differential pressure in a flow-rate microsensor [7.16].

membrane deflections with nitrogen gas. The feedback capacitor C_f is selected to produce the required output. For example, a 10 pF capacitor produces 0 to 5 V output over a sensing capacitance range of 5 to 14 pF. This design of flow microsensor suffers from temperature sensitivity and leakage currents. However, integrated thermodiodes and FET technology could be used to overcome these problems in applications where temperature fluctuations occur and low flow-rates are present.

7.4.4 Resonant bridge flow microsensors

The development of a flow microsensor based on the frequency shift of a resonating microbridge has also been reported [7.18]. Potential advantages of resonant microstructures are their high sensitivity, fast response and good stability. A $600 \times 200 \times 2.1$ μm stress-compensated silicon nitride microbridge has been fabricated by a front-side anisotropic etch of a groove, Figure 7.14. Thin-film phosphorus doped polysilicon resistors are embedded within the bridge for the thermal excitation and piezoresistive sensing of vibrations. The bridge is typically driven at a temperature elevation of 20°C with a base frequency of 85 kHz. A

Figure 7.14 Micrograph of an anisotropically-etched resonant microbridge flow-rate sensor [7.18].

frequency shift of 800 Hz was observed for a flow range of 0 to 10 ml/min. Consequently, resonant microstructures can be more sensitive than other types of flow sensors. However, the power consumption is somewhat larger to make the microbridge vibrate. The resonant frequency depends strongly upon the bridge mass. It is therefore essential that the microbridge is not contaminated by any particles in the fluid flow. This is clearly a potential drawback when the resonant microstructure sits within a real fluid.

7.5 Acceleration Microsensors

7.5.1 Basic considerations

Acceleration sensors or accelerometers are used to measure acceleration, vibration and mechanical shock (i.e. the result of an impulse load). Acceleration is the first derivative of velocity and second derivative of displacement. The relationships between velocity and displacement for a translational and rotational system are

$$a = \frac{dv}{dt} = \frac{d^2x}{dt^2} = \ddot{x}$$

$$\alpha_\theta = \frac{d\omega}{dt} = \frac{d^2\theta}{dt^2} = \ddot{\theta}$$

(7.16)

Although it is possible to compute the acceleration of an object from the output of a displacement or velocity sensor, most accelerometers use a sensing method in

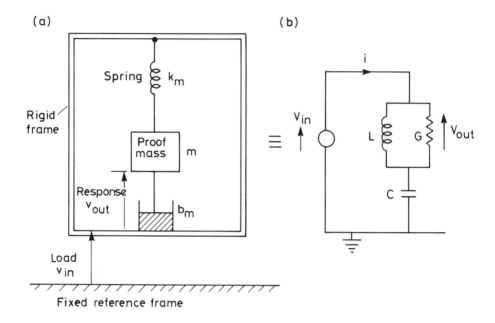

Figure 7.15 (a) General layout of an accelerometer and (b) its equivalent circuit.

which a proof mass is excited in a mass-spring-damper system. Figure 7.15(a) shows the basic sensing arrangement of an accelerometer.

The accelerometer is essentially a mass-spring-damper system in which the loading force $-m\ddot{x}_{in}$ drives a second order damped harmonic oscillator,

$$m\ddot{x}_{out} + b_m\dot{x}_{out} + k_m x_{out} = -m\ddot{x}_{in} \tag{7.17}$$

where x_{out} is the displacement of the proof mass relative to the rigid frame.

Under constant acceleration conditions, the displacement x_{out} is directly proportional to the input acceleration $\ddot{x}_{in}$ where,

$$x_{out} = \left(\frac{m}{k_m}\right)\ddot{x}_{in} \tag{7.18}$$

Steady-state conditions are reached in a time $t > (b_m/m)$, and so light damping and a heavy mass are required. A small spring constant k_m will ensure a good sensitivity. Under varying acceleration conditions, the dissipater plays a more important role. Figure 7.15(b) shows the electrical analogue of the mechanical system. The velocity load and response are equivalent to the voltages V_{in} and V_{out} shown in the circuit diagram. The velocity transfer function is thus,

$$\frac{V_{out}}{V_{in}} = \frac{Z_{LG}}{(X_C + Z_{LG})} \tag{7.19}$$

where Z_{LG} is the complex impedance of the parallel inductance-conductance network and X_C is the capacitive reactance.   Equation (7.19) has a rather cumbersome solution and so it is usually plotted as in Figure 7.16 showing the magnitude of the gain of the system against frequency for several damping factors ζ. The damping factor and natural frequency of vibration ω_0 of the mechanical and electrical systems are given by

$$\zeta = \frac{b_m}{2\sqrt{k_m m}} \equiv \frac{G}{2\sqrt{LC}}$$

$$\omega_0 = \sqrt{\frac{k_m}{m}} \equiv \sqrt{\frac{L}{C}}$$

(7.20)

This accelerometer can be run in different modes. Selecting a large mass and small spring-constant (at fixed damping) results in a low natural frequency and the proof mass is effectively stationary. The displacement x_{out} is π radians out of phase with the displacement x_{in} hence

$$x_{out} \approx -x_{in} \quad (\text{when } \omega > \omega_0, \ \omega_0 \text{ is low})$$

(7.21)

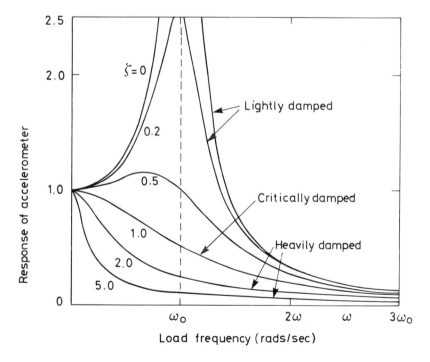

Figure 7.16 Response of an accelerometer using a second order damped system model. (Courtesy of Engineering department, University of Warwick, UK.)

Alternatively, selecting a small mass m and large spring-constant k_m results in a high natural frequency in which the proof mass tracks the frame,

$$x_{out} \approx 0 \quad (\text{when } \omega < \omega_0, \ \omega_0 \text{ is high}) \tag{7.22}$$

In this case, the displacement relative to a fixed reference frame can be measured.

Most accelerometers work on the principle of transducing the linear (or angular) displacement of the proof mass or spring into an electrical signal [7.7]. Various methods have been exploited in silicon microsensors and these include:

- piezoresistive/strain gauge;
- piezoelectric;
- capacitive;
- force balance (servo);
- acoustic.

Figure 7.17 illustrates some basic configurations of these different types of accelerometers (7.19). The advantages of using silicon microaccelerometers over

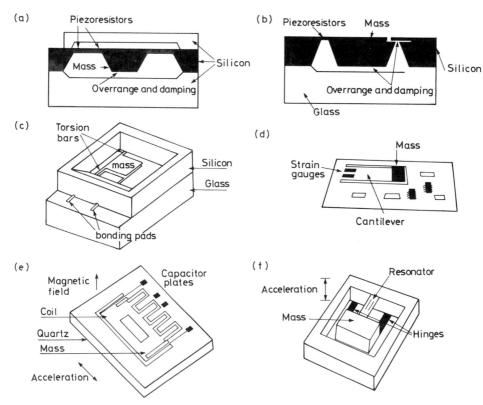

Figure 7.17 Some common types of accelerometers: (a) and (b) piezoresistive, (c) capacitive, (d) strain gauge, (e) force balance, and (f) resonant [7.19].

conventional accelerometers include

- low manufacturing cost;
- fast response time/wide dynamic range;
- low inertia mass;
- reliable and robust.

However, considerable care is needed when designing silicon microsensors because the spring-constant and damping coefficient are normally non-linear. This is usually obviated by a good geometric design to produce linear systems.

7.5.2 Piezoresistive microaccelerometers

The piezoresistive effect[2] in silicon is used for the measurement of various mechanical quantities, e.g. acceleration, pressure and force. Figure 7.18 shows the design of a microaccelerometer with a proof mass and cantilever beam micromachined out of single crystal silicon. The displacement of the proof-mass may be regarded as pure bending with any shear ignored. The maximum strain occurs at the cantilever base and this is measured by a piezoresistor, e.g. a boron-doped region of silicon.

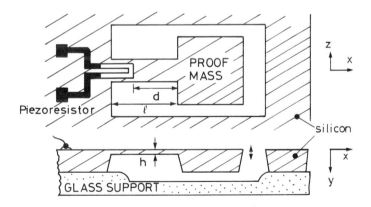

Figure 7.18 Basic design of a cantilever silicon microaccelerometer with piezoresistive read-out.

---

[2] Refer to § 7.8 for details.

The acceleration of the proof mass a_y (in the y-direction) can be related to the fractional change in resistance of the accelerometer [7.20] and is given by,

$$\frac{\Delta R}{R} = k_a a_y \tag{7.23}$$

The constant k_a of proportionality is approximately

$$k_a \approx \left[\tfrac{1}{2}(\Pi_{11} + \Pi_{12} + \Pi_{44}) - \Sigma_{12} - \tfrac{1}{2}\Sigma_{44}\right] \cdot \left[c_1 + m\left(\frac{(l+d)(c-h)}{2I_z}\right)\right] \tag{7.24}$$

where l, h and d are defined in Figure 7.18, and the proof mass m has a centre of gravity at $(l+c_1, c_2, c_3)$. c is the distance from the bottom of the beam to the centroid, while Π_{ij} and Σ_{ij} are the piezoresistive and elastic compliance coefficients, respectively. Provided that the beam width is much greater than the thickness, the acceleration in the third direction can be ignored. Thus, a mechanical design can ensure that the accelerometer behaves like the ideal elastic system.

The maximum sensitivity S_{max} is determined by the effective piezoresistive coefficient Π_{eff}, the fracture stress σ_f and geometry,

$$S_{max} = \frac{\Pi_{eff}\,\theta_f}{a_r}\left[\frac{2c_1 + l + d}{2(l + c_1)}\right] \tag{7.25}$$

where a_r is the range in acceleration.

Table 7.12 summarises the characteristics of a simple silicon microaccelerator first made in 1979 [7.20].

Table 7.12 Some characteristics of an early silicon microaccelerometer.

Size (mm)	Mass (mg)	Range (g)	Sensitivity $\frac{\Delta R}{R}/g$	Resonant frequency (kHz)	Non-linearity (% FSD)	Thermal sensitivity (%/°C)
$2 \times 3 \times 0.6$	20	± 200	2×10^{-4}	2.33	± 1	-0.2

Although this microaccelerometer shows a good linearity, it does have some drawbacks. In particular, the mechanical design yields a sensitivity to accelerations in the x and z directions and also the device has temperature-sensitivity mainly through the piezoresistors.

Considerable improvements have been made to the design of piezoresistive microaccelerometers. For example, a dual-beam piezoresistive microaccelerometer has been reported [7.21] which suppresses the torsional mode and has a built-in over-range stop. Consequently, it has a range of 0.2 to 100 g but can survive shocks of up to 1,000 g!

7.5.3 Capacitive and force balance (servo) microaccelerometers

The deflection of the proof-mass can also be measured capacitively. Figure 7.19 shows a cross-section of a doubly-fixed beam microaccelerometer with capacitive read-out from metal plates below and above the proof-mass.

The series capacitances in the network are given by,

$$C_1 = \frac{\varepsilon A}{x_1}; \quad C_2 = \frac{\varepsilon A}{x_2} \tag{7.26}$$

where ε is the permittivity of air and A is the area of the pick-up plate. As the displacement x_{out} (from equation 7.18) is proportional to the acceleration, the inverse capacitance of each capacitor is then proportional to the acceleration,

$$\frac{\delta C}{C} = \frac{x_{out}}{x} \propto \ddot{x} \tag{7.27}$$

Figure 7.20 shows the performance of a capacitive microaccelerometer using this arrangement over a range of ± 1 g [7.22]. The output capacitance varies from about 12 to 10 pF with a non-linearity of less than 0.5%, and the temperature sensitivity is only 20 μg/K.

For a small displacement of x_{out} of a dual element arrangement as shown in Figure 7.19, the ratio of capacitances is given by,

$$\frac{C_1}{C_2} = \frac{x_2}{x_1} \approx \frac{1+(x_{out}/x_1)}{1-(x_{out}/x_1)} \approx 1+\frac{2x_{out}}{x_1} \tag{7.28}$$

Thus, measuring the ratio of the capacitances eliminates the temperature dependence of the dielectric constant (and area).

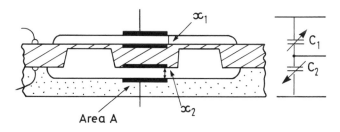

Figure 7.19 Cross-section of a silicon microaccelerometer with capacitive read-out.

Capacitive accelerometers generally have a higher stability, sensitivity and resolution than piezoresistive ones. The use of a passive half-bridge has the disadvantage of causing a displacement-dependent electrostatic force on the cantilever beam. A possible solution to this problem is to have a self-adjusting capacitor bridge in which the applied voltage is modulated to achieve zero output voltage from the tap point. Such a force balance microaccelerometer has been realised [7.23] with integrated CMOS circuitry.

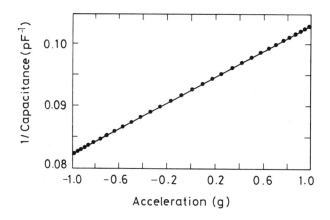

Figure 7.20 Performance of a capacitive microaccelerometer [7.22].

In essence, the proof-mass is held in a fixed position by applying a voltage to balance the electrostatic force. As the force balancing voltage is proportional to the electrostatic force, maintaining equilibrium results in a voltage V_{fb} proportional to acceleration,

$$V_{fb} \propto k\ddot{x} \tag{7.29}$$

The bandwidth of the force-balance microaccelerometer is limited by the performance of the feedback circuitry and the non-linear displacement terms disappear to give excellent linearity. For example, a working range of ± 0.1 g with a resolution of 1 μg at 1 Hz and temperature coefficient of about 30 $\mu g/K$ are achievable [7.23].

7.5.4 Comparison of performance

Conventional accelerometers often use a piezoelectric crystal element in which charge is induced by the inertial force; such acoustic accelerometers are reviewed in [7.24]. Table 7.13 compares these to the performance of microaccelerometers and states some of their applications. A force-balance accelerometer may be controlled electrostatically; however, an electromagnetic version has also been reported. The low cost of some types of microaccelerometer, e.g. piezoresistive and capacitive, has led to their successful application in the automobile industry. Such applications include the deployment of air-bags in cars to protect against personal injury in a road accident.

Finally, the resonance of a microaccelerometer can be used to detect accelerations. In a resonant mode, the frequency of vibration depends upon the tension in the cantilever supports. Under acceleration, the proof-mass is displaced causing the tension and thus spring-constant in the cantilever beam to change. This in turn results in the resonant frequency shifting. The technique is more commonly used in force and pressure sensors (see §7.6).

Table 7.13 Comparison of microaccelerometers.

Parameter	Piezoelectric	Piezoresistive (displacement)	Capacitive (displacement)	Force-balance
Size	Large	Small	Small	Medium, needs ICs
Precision	Good	Good	Good	Very good
Linearity	Poor	Good	Good	Very good
Resolution	Good	Good	Good	Very good
Range (g)	250	±100	±100	±0.1, ±1.0
Cost	Medium	Low	Low	Medium/High
Application	Manufacturing Military	Automobile Manufacturing General purpose	Automotive Manufacturing General purpose	Navigation Military Aerospace

7.6 Force, Pressure and Strain Microsensors

7.6.1 Basic considerations

The intrinsic properties of materials are often exploited in the design of force, pressure and strain microsensors. Principally, the piezoelectric and piezoresistive effects are utilised in micromechanical devices.

When a force F_q is applied to a plate of piezoelectric material, a charge q is induced onto the surface. The charge induced is given by,

$$q = \Xi F_q \qquad (7.30)$$

where Ξ (in units of coulombs per newton) is the piezoelectric coefficient and is strictly a tensor with terms dependent on the lattice orientation. Silicon has a centrosymmetric crystal lattice structure and so does not exhibit piezoelectricity. It is therefore necessary to deposit piezoelectric layers onto silicon micromechanical structures. ZnO is a common material to use although $BaTiO_3$ or $PbZrTiO_3$ (PZT) have higher piezoelectric coefficients. Table 7.14 shows the piezoelectric coefficients for some common materials.

Table 7.14 Piezoelectric coefficients of materials at 300 K.

Material	Type	Form	Coefficient, Ξ_{33} (pC/N)	Permittivity, ε_r
Quartz (X-cut)	Glass	Bulk	2.33	4.0
PVDF	Polymer	Film	1.59	-
P(VDF-TrFE)	Polymer	Film	18.0	6.2
ZnO	Ceramic	Bulk	11.7	9.0
ZnO	Ceramic	Film	12.4	10.3
$BaTiO_3$	Ceramic	Bulk	190	4,100
PZT	Ceramic	Bulk	370	300- 3,000

In the piezoresistive effect, the relative change in resistance or resistivity is proportional to the mechanical stress σ_m. In matrix notation

$$\frac{\Delta R_i}{R_i} = \Pi_{ij}\, \sigma_{mj} \qquad (7.31)$$

where Π_{ij} is the ijth piezoresistive coefficient of the material matrix.

In silicon, there are only three non-zero piezoresistive coefficients, namely Π_{11}, Π_{12} and Π_{44} due to the centrosymmetric structure. Piezoresistors can be readily fabricated by diffusing n- or p-type materials into the bulk silicon crystal. Table 7.15 shows the piezoresistive coefficients for n-type and p-type silicon [7.25].

Table 7.15 Piezoresistive coefficients of silicon [7.25].

Doping	Resistivity (Ωcm)	Π_{11}	Π_{12}	Π_{44}
n-type	+11.7	−102.2	+53.4	−13.6
p-type	+7.8	+6.6	−1.12	+138.1

The fractional change in resistance of a piezoresistor is related to the parallel and transverse piezoresistive coefficients, Π_p and Π_t, and stress σ_p and σ_t:

$$\frac{\Delta R}{R} = \Pi_p\, \sigma_p + \Pi_t\, \sigma_t \qquad (7.32)$$

The exact fractional change in resistance depends upon the crystal orientation used, e.g. (100), (110) or (111) crystal planes.

It is often more convenient to express the effect in terms of strain ε_m rather than stress. For a thin film material of length l, the gauge factor K_{gf} is defined as,

$$K_{gf} = \frac{\Delta R / R}{\Delta l / l} = \frac{\Delta R / R}{\varepsilon_m} \tag{7.33}$$

The gauge factor is a measure of the strain sensitivity of a material. Large gauge factors are obtained for n-type silicon in the [100] direction ($K_{gf} = -153$) and p-type silicon in the [111] direction ($K_{gf} = +173$). The resistance change can be found from the sum of the longitudinal and perpendicular gauge factors and strains,

$$\frac{\Delta R}{R} = K_p \varepsilon_p + K_t \varepsilon_t \tag{7.34}$$

The gauge factors depend not only on the doping level but also on the temperature. Figure 7.21 shows the temperature sensitivity of the gauge factor to temperature and doping level in p-type silicon [7.26].

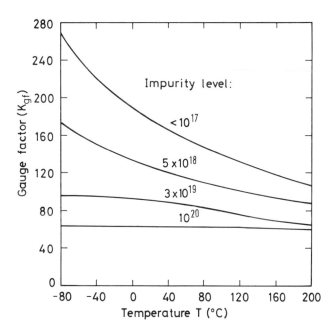

Figure 7.21 Effect of temperature on gauge factor of a p-type silicon piezoresistor at various doping levels [7.26].

7.6.2 Force microsensors

Force microsensors generally use a sensing element that converts the applied force into the deformation of an elastic element. For example, a cantilever beam is displaced by the application of a static load and the displacement can be measured capacitively or through the strain.

A micromechanical stylus-force gauge has been fabricated with a range of 1 to 500 mN with an accuracy of ±1 mN. The deflection of a 100 μm square plate is measured using polysilicon piezoresistors and a half-bridge arrangement [7.27]. A possible application for these microsensors is in scanning force microscopy where a small stylus structure is used to measure the surface profile of a material. Figure 7.22 shows an SEM photograph of a simple beam cantilever that is used in Atomic Force Microscopes (AFM) [7.28].

7.6.3 Stress-sensitive electronic devices

Mechanical stress is the force per unit area which when acting on a surface is called pressure. When a stress is applied to a body (within its elastic limit) a linearly-related strain is produced.

The energy levels of the valence band of silicon split and become non-degenerate under strain. Table 7.16 shows the changes in band-gap E_g and degenerate X_1 and Γ_{25} levels under uniaxial stress in the <100>, <111> and <011> directions [7.29].

Figure 7.22 Micrograph of a silicon microcantilever above an IC [7.28].

Table 7.16 Effect of stress on silicon band structure [7.29].

Parameter:	% Change under uniaxial stress in direction		
	<100>	<111>	<011>
Band-gap	−10.83	−10.91	−10.23
Top of valence band, $\Gamma_{25'}$	+5.17/−1.73	+4.88/−3.45	+4.03/−2.58
Bottom of conduction band,			
<100>	−5.66	+7.25/−6.03	+13.74/−6.20
<010> / <001>	+3.76	+7.25/−6.03	−0.95

The major consequence of a change in band structure is to modify the carrier concentration. In doped semiconductor materials, the band-gap shift ΔE_g produces a change in minority carrier concentration of

$$p_\sigma/p_0 = \exp(\Delta E_g/kT) \qquad (n\text{-type})$$

$$n_\sigma/n_0 = \exp(\Delta E_g/kT) \qquad (p\text{-type})$$

$$(7.35)$$

The shift in band-gap ΔE_g is approximately proportional to the pressure P,

$$\Delta E_g = k_p P \qquad (7.36)$$

where the constant k_p is typically 10^{-10} eV/Pa. Thus, a large pressure is required in order to produce a significant change (> 1 GPa!).

From basic semiconductor theory, the forward current flowing through an n-p diode can be related to the applied stress σ_m (or pressure) by

$$I_\sigma/I_0 = \exp(\Delta E_g/eV_f) = \exp\left(k_p P/eV_f\right) \text{ or } \exp\left(k_p \sigma_m/eV_f\right) \qquad (7.37)$$

where I_0 is a current constant given by the ratio of applied area to total area.

Figure 7.23 shows the diode current I_σ as a function of forward voltage at various stress levels [7.30]. The stress has been applied by a stylus tip and so a deviation from the ideal exponential curve is observed and depends upon the applied area of stress and total device area. Stylus-based force diode sensors have not proved successful because of their low sensitivity and a vulnerability to damage by shock.

Nevertheless studies have been made on the force and pressure sensitivity of bipolar transistors and MOSFETs. The effect of stress on the emitter-base junctions of a bipolar transistor is to decrease the current gain β_σ (I_b/I_c) or reduce the base-emitter voltage V_{be} at constant collector current I_c. The collector and base currents are approximately given by (applied area $\approx$ total junction area) [7.29],

$$\frac{I_c}{I_0} \approx I_s \exp(k_p P/eV_f).\left[\exp(V_{be}/V_f) - 1\right]$$

(7.38)

$$\frac{I_b}{I_0} \approx I_{b1} \exp(k_p P/eV_f) + I_{b2} \exp(k_p P/eV_f)$$

where I_s is the saturation current, I_{b1} the recombination current, and I_{b2} the current from injection of holes from base to emitter.

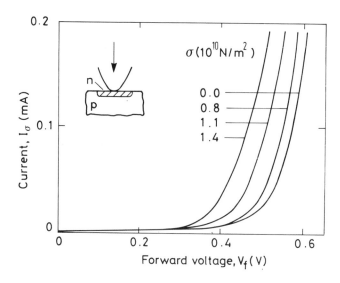

Figure 7.23 Effect of stress upon the characteristics of a silicon *n-p* diode.

Figure 7.24 shows the effect of a force F (and so stress), applied by a 32 μm radius tip, on the current gain and base-emitter voltages of a transistor. The corresponding stress is also shown. The change in base-emitter voltage (at constant current) is proportional to the force and is perhaps a more practical parameter. A low cost force sensor based on a bipolar transistor has been proposed [7.31]. More recently, MOSFET force sensors are also being proposed, in which the stress reduces the carrier mobility and hence the saturation current. However, both types of device suffer from transduction characteristics that depend strongly upon process physical parameters and a complex design.

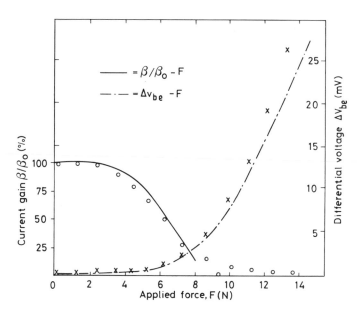

Figure 7.24 Effect of stress (force) upon the characteristics of a bipolar transistor.

7.6.4 Pressure microsensors

Some notable success has been achieved in the use of micromachining to produce silicon pressure sensors. The simplest design is that consisting of a thin silicon diaphragm with piezoresistors diffused into the surface that measure its deflection [7.32]. Figure 7.25(a) shows the general arrangement of a silicon diaphragm pressure sensor with piezoresistive pick-up. Single-crystal silicon is an excellent material for pressure sensors because practically no creep or hysteresis occurs. Diffusing two p-type resistors into an n-type silicon diaphragm of surface orientation (100) results in the piezoresistive coefficients (from equation (7.21)) becoming,

$$\Pi_{11} = -\Pi_p = \tfrac{1}{2}\Pi_{44} \tag{7.39}$$

where the piezoresistive constant Π_{44} is typically +138.1 pC/N (see Table 7.15). Using a half-bridge arrangement as shown with $R_1=R_2$ at zero stress and an input voltage of V_{ref} gives

$$V_{out} = \frac{1}{2}\left[1+\frac{1}{2}\Pi_{44}(\sigma_{1y}-\sigma_{1x})V_{ref}\right] \tag{7.40}$$

where σ_{1x} and σ_{1y} are the stresses induced by the applied pressure P.   The sensitivity is then given by

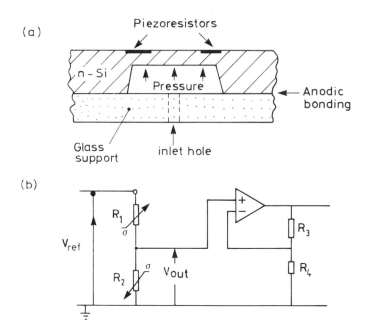

Figure 7.25 (a) Schematic arrangement of a silicon diaphragm pressure sensor, and (b) read-out circuit.

$$S = \frac{1}{V_{ref}} \frac{\partial V_{out}}{\partial P} = \frac{1}{4} \Pi_{44} \frac{\partial(\sigma_{1y} - \sigma_{1x})}{\partial P} \qquad (7.41)$$

The pressure range and sensitivity depend upon the geometry of the silicon diaphragm, i.e. cross-sectional area and thickness. The signal conditioning usually takes the form of an op-amp circuit, e.g. Figure 7.25(b) with the gain of $(1 + R_3/R_4)$. Pressure sensors with integrated monolithic bipolar circuitry [7.33] and bipolar I^2L [7.34] ring-oscillator circuits have been reported. On- or off-chip temperature compensation is required as the piezoresistive constant is temperature dependent.

Silicon piezoresistive sensors are now widely used because of their low price and excellent performance. Recently, an integrated device suitable for air brakes of automobiles has been reported [7.34]. A 0.8 MPa pressure sensor was fabricated with a linearity of ±20%. Dynamic ranges of a few hundred millibar to a few hundred bar are quite reliable.

Biomedical application of piezoresistive pressure sensors in catheter tubes is also possible. However, burying of the piezoresistors helps ameliorate the effect of surface contamination and thus base-line drift [7.35].

Capacitive read-out from silicon diaphragm pressure sensors is also used although such devices are generally more expensive to fabricate.

A more radical device is one using a micromachined resonant structure attached to the diaphragm whose frequency shifts with pressure. Figure 7.26 shows the basic arrangement of a torsional resonant pressure sensor. The

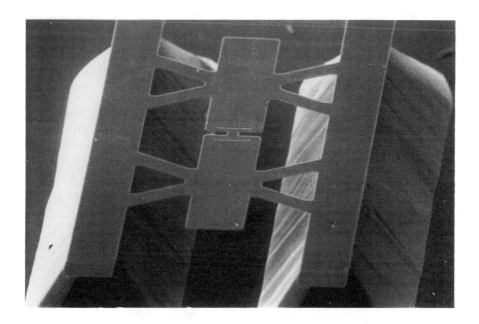

Figure 7.26 Resonant microstructure for silicon pressure sensor [7.36].

deflection in the diaphragm puts the proof-mass supports under tension. This in turn modifies the torsional spring-constant and thus the resonant frequency of the coupled oscillators. Using coupled oscillators helps improve the mechanical stability. Various methods have been reported on the excitation/detection of the resonant structure. For example, capacitive excitation and pick-up has been reported [7.36] with careful calibration of the non-linear response and temperature-sensitivity. The advantage of the resonant structure over a static structure is the improved accuracy and lower hysteresis. More recently, interest has been shown in optically excited resonant pressure sensors [7.37]. Possible applications include those in which an electrical signal is potentially hazardous.

7.6.5 Strain microsensors

Resistive strain sensors are widely used as a low cost and simple way to measure forces, torques, pressures, displacements etc. The resistance R of a strip of material of length l, cross-sectional area A and resistivity ρ is

$$R = \frac{\rho l}{A} \tag{7.42}$$

Applying the total differential theorem to equation (7.42) leads to

$$\frac{\Delta R}{R} = \frac{\Delta l}{l} - \frac{\Delta A}{A} + \frac{\Delta \rho}{\rho} \tag{7.43}$$

The fractional change in area for a homogeneous material is approximately related to the fractional change in length via Poisson's ratio v_m hence

$$\frac{\Delta R}{R} \approx (1 + 2v_m)\frac{\Delta l}{l} + \frac{\Delta \rho}{\rho} \qquad (7.44)$$

So the equation which describes the fractional change in resistance has a dimensional term which depends upon the length $\Delta l / l$ and a piezoresistive term which depends upon the bulk property.

Metal foil strain-gauges consist of a thin metal foil with an insulating support. Figure 7.27 shows some examples of the design of metal foil strain-gauges. The piezoresistive effect in metals is small and so the resistance change and hence gauge factor K_{gf} is given by

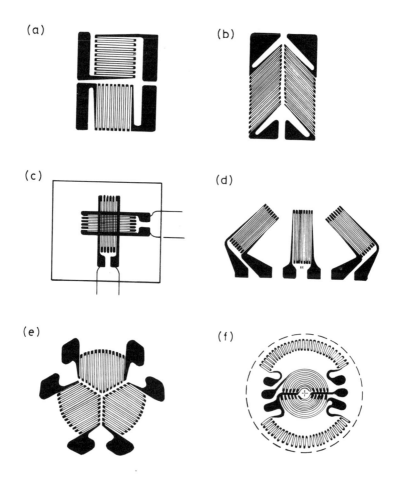

Figure 7.27 Six designs of thin foil strain-gauges.

$$\frac{\Delta R}{R} \approx (1+2v_m)\frac{\Delta l}{l} \quad \text{so} \quad K_{gf} = \frac{\Delta R/R}{\Delta l/l} = (1+2v_m) \qquad (7.45)$$

The gauge factor is low, because Poisson's ratio is typically ¼ to ½, and typically takes a value of 2.0 for a copper-nickel alloy foil. Strains of between 2% and 4% can be measured by adhering the foil gauge (down to about 10^{-5}) onto the surface of the material under stress. The composition of the alloy can be modified so that its linear expansion matches that of the material under test, e.g. aluminium or mild steel. Typically gauge resistances are ~100 Ω and cost only a few pounds.

Higher gauge factors have been reported for thick-film printed composite materials [7.38] but these materials are not as compatible with IC technology as metal films.

In piezoresistive strain gauges, the bulk term tends to dominate and so equation (7.44) may be rewritten as,

$$\frac{\Delta R}{R} \approx \frac{\Delta \rho}{\rho} \quad \text{as} \quad \frac{\Delta l}{l} \to 0 \quad \text{(Piezoresistive devices)} \qquad (7.46)$$

Thus thin strain-sensitive silicon films (down to about 10^{-6}) are commonly used in many mechanical microsensors (see previous sections). Large gauge factors are observed for silicon p- and n-type gauges. Polycrystalline silicon may also be used as a strain gauge material but its gauge factor is smaller, lying between that of a metal foil (~2) and a p-type silicon device (~150) [7.39]. The exact gauge factor depends upon the orientation and size of the grains but its wide operating temperature and stable output make it attractive in certain applications.

Finally, it is possible to measure low levels of strain in a piezoelectric crystal. The resonant frequency of a surface acoustic wave device is strain dependent, and reported as [7.40]

$$f \approx f_0(1-1.28\varepsilon_m) \qquad (7.47)$$

A device with a basic resonant frequency f_0 of 140.2 MHz yielded a strain sensitivity of 1.28 and a short-term resolution of 0.01 microstrain!

7.7 Mass Microsensors

7.7.1 Basic considerations

The measurement of mass is generally called *gravimetry*, but clearly conventional scales or spring balances are not regarded here as mass microsensors. A force microsensor (§7.6.2) can be regarded as a mass sensor when the force is applied by a dead-weight. Force microsensors generally measure large forces through the deflection of a silicon or some other type of structure. However, there is another class of mass sensors which are called microbalances and measure very small masses. There are two main types of microbalances based on either piezoelectric mass or Surface Acoustic Wave (SAW) devices.

7.7.2 *Piezoelectric mass microsensors*

A piezoelectric crystal can be made to vibrate at its natural frequency and may be modelled as a second order harmonic oscillator (§7.5.1). For a crystal of mass m and thickness d, the shear-mode has a wavelength of $2d$ at resonance and the fractional change in frequency is related to the mass and thickness by

$$\frac{\Delta f_m}{f_m} = -\frac{\Delta m}{m} = -\frac{\Delta d}{d} \qquad (7.48)$$

The addition of a small foreign mass m_f over the crystal surface may be treated as an equivalent mass and will shift the resonant frequency,

$$\frac{\Delta f_m}{f_m} = -\frac{m_f}{m} \qquad (7.49)$$

Assuming that the deposited material has a uniform density, then the foreign mass is given in terms of the acoustic shear velocity v_m by

$$m_f \approx -\frac{\Delta f_m}{2 f_m^2} \rho_m v_m \quad \text{or} \quad \Delta f_m = -\Lambda_f m_f \qquad (7.50)$$

where Λ_f is a crystal calibration constant. Ideally, the acoustic impedance of the foreign material Z_f should match that of the piezoelectric material. There is a variety of materials which can be used in mass sensors as listed in Table 7.17.

Table 7.17 Acoustic impedances of materials used in mass sensors.

Material	Impedance (10^6 kg s/m^2)	Material	Impedance (10^6 kg s/m^2)
Pt	36.1	Ag	16.7
Cr	29.0	Si	12.4
Ni	26.7	In	10.5
Al$_2$O$_3$	24.6	SiO$_2$ (quartz)	8.27
Pd	24.6	Al	8.22
Au	23.2	C (graphite)	2.71
Cu	20.3		

A quartz microbalance is a very sensitive and stable microsensor. For example, an AT-cut quartz (ρ_m=2,650 kg/m^3 and v_m=3,340 m/s) plate with a resonance at 5 MHz has a typical crystal calibration constant of 5.65 MHz cm^2/kg. Consequently, a 1 Hz shift represents an added weight of 17.7 ng per centimetre squared!

The deposition of a chemically-sensitive material onto the surface of the piezoelectric crystal leads to a chemical sensor. The coating usually takes the form of ultra-thin biological films or thin organic films (§9.7) using, for example, a Langmuir-Blodgett technique (§4.5.3).

7.7.3 SAWR microsensors

The shear waves in piezoelectric crystals travel through the bulk of the material. In contrast, in Surface Acoustic Wave Resonant (SAWR) devices the acoustic wave travels only near the surface of the crystal. An advantage of the SAWR microbalance is that the sensing part of the device is separate from the piezoelectric transducer and receivers [7.41]. This is illustrated in Figure 7.28(a) where the SAW is generated by an R.F. oscillator (T) and the phase shift $\Delta\Phi$ (or frequency shift Δf) is measured by a receiver (R). Most SAW devices employ a dual arrangement with a delay line and reference line, see Figure 7.28(b).

The exact principles of the SAW device are rather more complex than a piezoelectric device. The output V_{out} is governed by the reference and sensing

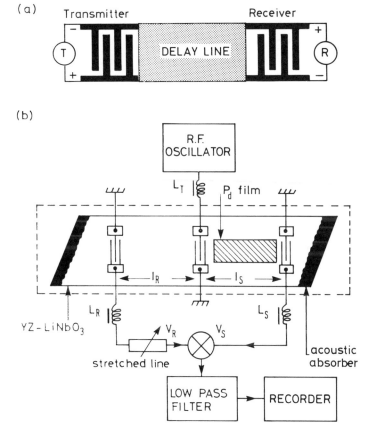

Figure 7.28 Schematic layout of (a) a single delay line, and (b) dual element SAWR mass microsensors.

phase shifts respectively,

$$V_{out} \propto \sin(\Phi_r - \Phi_s) \tag{7.51}$$

where $\Phi_r = 2\pi f_0 l_r / v_r$ (+ delay) and $\Phi_s = 2\pi f_0 l_s / v_s$. The additional mass changes the acoustic velocity and thus modifies the phase shift $(\Phi_r - \Phi_s)$. The maximum output at zero mass is obtained when $(\Phi_r - \Phi_s)$ is $n\pi/2$ with the sensor usually excited in the fundamental mode (n=1).

The SAW sensor substrate is usually made of ST-cut quartz or $LiNbO_3$. This type of device is commonly used in chemical sensing where the reference material absorbs/desorbs a chemical species (§9.7).

7.8 Other Mechanical Measurands

7.8.1 Introduction

There are a variety of other mechanical parameters which be measured using microsensors. A few of them are briefly discussed here, namely, roughness and shape, density and viscosity, and sound waves.

7.8.2 Roughness and shape

Conventional gauging techniques, such as optical interferometry, are used to measure the profile of a surface. Recently, there has been considerable development of microprobes capable of resolving surface detail down to the submicron and even atomic level [7.42]. This has been achieved through two designs, namely the Scanning Tunnelling Microscope (STM) and Atomic Force Microscope (AFM). Both microscopes make use of a micromechanical probe to scan the surface but the STM measures the contours of constant electron density by maintaining a constant tunnelling current whereas the AFM measures the mechanical deflection of a probe by capacitive or other means, Figure 7.29. The interest in a micromechanical sense, is that the measurement systems rely upon a microprobe with a very sharp tip. Consequently, some work has been put into the use of silicon micromachining techniques to fabricate precise microprobes [7.28]. A roughness parameter R_q can be calculated from the surface height profile $h(x)$ and is defined as,

$$R_q = \sqrt{\frac{1}{l} \int_0^l h^2(x) dx} \tag{7.52}$$

where l is the gauge length over which the surface has been profiled. R_q is a measure of the RMS height and so characterises the surface finish. However, it is not always simple to relate the output from a microprobe to the mechanical measure of surface roughness. This is because the microprobe can interfere with the measurement process and because of the fact that there is no distinct surface at the microscopic level!

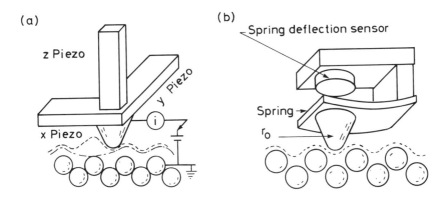

Figure 7.29   Schematic layout of (a) an STM and (b) an AFM for profiling surfaces at the submicron level.

Shape may be regarded as the external profile of an object. There are many conventional techniques used to measure the shape of an object such as laser gauging and optical holography. Some discussion is made later on the use of optical arrays and tactile microsensor arrays to measure the shape by profiles and touch (Chapter 12). However, perhaps the simplest shape sensor is a particle filter which sets an upper limit to the particle diameter. Figure 7.30 shows the cross-section of a silicon micromachined microfilter [7.43]. The membrane has been etched from single crystal silicon using a boron etch-stop technique followed by SiO_2 undercut etching. Filters have been made which only allow particles of less than 50 nm diameter through and have potential application in microbiology for filtration of DNA molecules.

7.8.3 Viscosity and density

Viscosity is a property of a fluid and is a measure of its tendency to resist motion. For two parallel layers in the flow direction, the viscosity η_m is the proportionality constant between the viscous force F_v and velocity gradient dv/dx between the layers, hence

$$F_v = \eta_m \frac{dv}{dx}$$

(7.53)

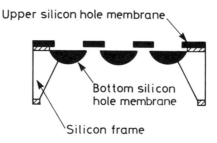

Figure 7.30   Schematic diagram of the cross-section of a mechanical microfilter [7.43]; also see Figure 12.2.

For laminar flow, the viscosity enters Poiseuille's equation describing the flow-rate Q_v through a pipe, namely

$$Q_v = \frac{\pi a^4}{\eta_m} \left(\frac{\Delta P}{\Delta l} \right)$$

(7.54)

where a is the radius of the pipe and $\Delta P / \Delta l$ is the pressure dropped across its length l. Consequently, the viscosity of a fluid can be determined by measuring the viscous forces, flow-rates and pressure drops. Clearly this could be achieved using micromechanical flow-rate and pressure sensors. However, another method has been reported recently that makes use of an ultrasonic microbridge to measure both viscosity and density [7.44]. Figure 7.31 shows the design of the sensor. A thin ZnO layer (0.7 μm) lies on top of a 2.0 μm Si_3N_4 membrane onto which aluminium interdigital transducers have been deposited. Measurement of the wave transmission loss L and frequency response of the delay-line correlate to the fluid viscosity and density, Figure 7.32.

7.8.4 *Acoustic and ultrasonic microsensors*

Acoustic waves are longitudinal pressure waves that are audible to the human ear. Audio or sonic waves cover a frequency range of about 15 Hz to 20 kHz. Infrasound lies below this range and ultrasound above it. The intensity and frequency of sound is measured conventionally using microphones in air or hydrophones in liquid, especially in water. Sound is sensed by pressure-sensing elements. Condenser microphones are capacitive sound-pressure sensors while piezoelectric microphones and most hydrophones use pressure-sensitive ceramic or quartz crystals [7.7].

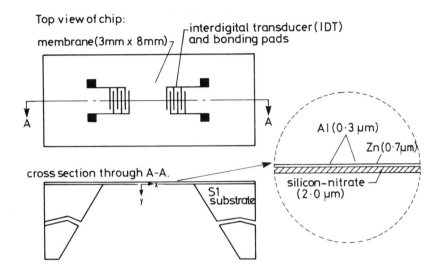

Figure 7.31 Schematic diagram of an ultrasonic viscosity microsensor [7.44].

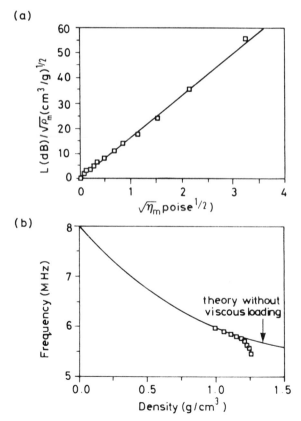

Figure 7.32 Effect of viscosity and density on an ultrasonic microsensor [7.44].

Although materials technology has improved the efficiency of acoustic sensors and hydrophones, for example through the use of piezoplastics and PZT-polymer composites [7.45], this book's interest lies in the application of silicon processing techniques to acoustic devices. The fabrication of a ZnO layer sputtered onto a thin etched silicon diaphragm was reported in 1983 [7.46]. Figure 7.33 shows the structure of the acoustic microsensor or microphone. The acoustic sensitivity was 250 μV/Pa with a dynamic range of 0.1 Hz to 10 kHz and power consumption of below 40 μW. An integrated MOS amplifier and annular electrode structure helped reduce parasitic capacitance and temperature sensitivity. More recently, a microphone has been fabricated using only silicon in which the deflection of a diaphragm is measured capacitively [7.47]. The acoustic sensitivity was 4.2 fF/Pa or 15 mV/Pa. A dynamic range of about 100 Hz to 20 kHz was achieved with the sensitivity highest for a 5 μm diaphragm thickness (Figure 7.34).

A similar mechanical principle can be employed to measure ultrasound with a microsensor. However, a $PbTiO_3$ active layer is required to provide significant coupling to the micromechanical structure, e.g. a cantilever beam. Resonant frequencies of Si and SiO_2 between 10 and 200 kHz are realisable by varying the cantilever length [7.48]. It is much cheaper to use this type of sensing principle than conventional optical techniques, such as interferometry [7.49].

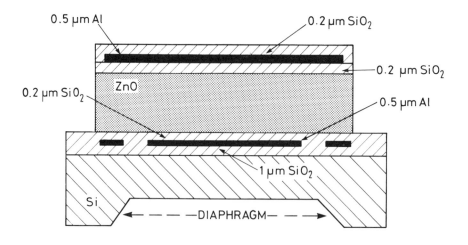

Figure 7.33 Schematic diagram of a ZnO on silicon microphone [7.46].

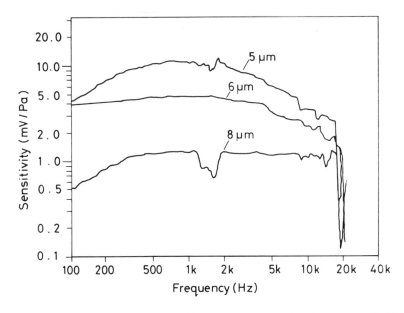

Figure 7.34 Dynamic response of a capacitive acoustic microsensor [7.47].

Suggested further reading

Readers are referred to the following texts that provide some background information for this chapter on mechanical microsensors:

Case J and Chilver AH: *Strength of Materials* (1971). Published by Arnold, London. ISBN 0-7131-3244-2. (404 pages. A good introduction to mechanics)

Young WC: *Roark's Formulas for Stress and Strain* (1989). Published by McGraw-Hill, New York. ISBN 0-07-100373-8. (763 pages. A reference book for behaviour of mechanical structures)

Göpel W (ed.): *Sensors: A Comprehensive Survey, Vol. 7: Mechanical Sensors* (1993). Published by VCH, Germany. ISBN 3-527267735 (600 pages. An overview of mechanical sensors)

References

7.1   Dubay CL (1984) A transduction path method of solid state sensor analysis and investigation. *MS thesis, Naval Postgraduate School, Monterey, CA, USA*, 166 pp.

7.2   White RM (1987) A sensor classification scheme. *IEEE Trans. on UFFC*, **34**, 124-126.

7.3   Nye JF (1979) *Physical Properties of Crystals*, Oxford University Press, UK, 322 pp.

7.4   Petersen KE (1982) Silicon as a mechanical material. *Proc. IEEE*, **70**, 420-457.

7.5   van Vlack LH (1964) *Elements of Material Science and Engineering*, Addison Wesley, Reading, MA, USA, p.163.

7.6   Howe RT (1980) Resonant microsensors. *Technical Digest, Transducer '87, 4th Int. Conf. on Solid-state Sensors and Actuators, Tokyo, Japan*, pp. 834-849.

7.7   Norton HN (1989) *Handbook of Transducers*, Prentice-Hall Inc., New Jersey, USA, 554 pp.

7.8   Bowen DK (1991) Calibration of linear transducers by X-ray interferometry, in *From Instrumentation to Nanotechnology* (eds. J.W. Gardner and H.T. Hingle), Gordon and Breach Science Publishers, Philadelphia, USA, Ch. 16, pp. 319-331.

7.9   Kato H, Kojima M, Okumura Y and Ozaki A (1990) Photoelectric inclination sensor. *Sensors and Actuators*, **A21-A23**, 289-292.

7.10 Canali C, de Cicco G, Morten B, Prudenziati M and Taroni A (1981/2) An ultrasonic proximity sensor operating in air. *Sensors and Actuators*, **2**, 97-103.

7.11 Magori V and Walker H (1987) Ultrasonic presence sensors with wide range and high resolution. *IEEE Trans. on UFFC*, **34**, 202-211.

7.12 Johnson RG and Higashi RE (1987) A highly sensitive chip microtransducer for air flow and differential pressure sensing application. *Sensors and Actuators*, **11**, 63-72.

7.13 Stemme GN (1986) A monolithic gas flow sensor with polyimide as thermal insulator. *IEEE Trans. on Electron Devices*, **33**, 1470-1474.

7.14 Oudheisen BW and Huijsing JH (1990) An electronic wind meter based on a silicon flow sensor. *Sensors and Actuators*, **A21-A24**, 420-424.

7.15 *Honeywell Technical Note on Mass Flow Microbridge*.

7.16 Cho ST and Wise KT (1993) A high-performance microflowmeter with built-in self-test. *Sensors and Actuators A*, **36**, 47-56.

7.17 Tabor D (1979) *Gases, Liquids and Solids*, Cambridge University Press, Cambridge, UK, p. 265.

7.18 Bouwstra S, Legtenberg R, Tilmans HAC and Elwenspeek M (1990) Resonating microbridge mass flow sensor. *Sensors and Actuators*, **A21-A23**, 332-335.

7.19 MacDonald GA (1990) A review of low cost accelerometers for vehicle dynamics. *Sensors and Actuators*, **A21-A23**, 303-307.

7.20 Roylance LM and Angeli JB (1979) A batch-fabricated silicon accelerometer. *IEEE Trans. on Electron Devices*, **26**, 1911-1917.

7.21 Barth PW, Pourahmadi F, Mayer R, Poydock J and Petersen K (1988) A monolithic silicon accelerometer with integral air damping and over-range protection. *IEEE Solid-state Sensors and Actuators Workshop, Hilton Head Island, SC*, pp. 114-116.

7.22 Seidel H, Riedel H, Kolbeck R, Mück G, Kupke W and Königer M (1990) Capacitive silicon accelerometer with highly symmetrical design. *Sensors and Actuators*, **A21-A23**, 312-315.

7.23 Rudolf F, Jornod A, Bergqvist J and Leuthold H (1990) Precision accelerometers with µg resolution. *Sensors and Actuators*, **A21-A23**, 297-302.

7.24 Motamedi M (1987) Acoustic accelerometers. *IEEE Trans. on UFFC*, **34**, 237-242.

7.25 Smith CS (1954) Piezoelectric effect in germanium and silicon. *Phys. Rev.*, **94**, 42-49.

7.26 Mason WP (1969) Use of solid-state transducers in mechanics and acoustics. *J. Aud. Engineering Soc.*, **17**, 506-511.

7.27 Tai Y-C and Muller RS (1990) Integrated stylus-force gauge. *Sensors and Actuators*, **A21-A23**, 410-413.

7.28 Brugger J, Buser RA and de Rooij NF (1992) Silicon cantilevers and tips for scanning force microscopy. *Sensors and Actuators A*, **34**, 193-200.

7.29 Longwell TF (1968) An experimental semiconductor microphone. *Automotive Elect. Tech. J.*, **11**(3), 109-116.

7.30 Clark SK and Wise KD (1979) Pressure sensitivity in anisotropically etched thin-diaphragm pressure sensors. *IEEE Trans. on Electron Devices*, **26**, 1887-1896.

7.31 Sansen W, Vandeboo P and Puers B (1982/3) A force transducer based on stress effects in bipolar transistors. *Sensors and Actuators*, **3**, 343-354.

7.32 Borky JM and Wise KD (1979) Integrated signal conditioning for silicon pressure sensors. *IEEE Trans. on Electron Devices*, **26**, 1906-1910.

7.33 Dorey AP and French PJ (1983) Frequency output piezoresistive pressure sensors. *Proc. Solid-State Transducers '83*, p.141.

7.34 Ansermet S, Otter D, Craddock RW and Dancaster JL (1990) Cooperative development of a piezoresistive pressure sensor with integrated signal conditioning for automotive and industrial application. *Sensors and Actuators*, **A21-A23**, 79-83.

7.35 Esashi M, Komatsu H and Matsuo T (1983) Biomedical pressure sensor using buried piezoresistors. *Sensors and Actuators*, **4**, 537-544.

7.36 Greenwood JC (1988) Silicon in mechanical sensors. *J. Phys. E: Sci. Instrum.*, **21**, 1114-1128.

7.37 Thornton KEB, Uttamchandani D and Culshaw B (1990) A sensitive optically excited resonator pressure sensor. *Sensors and Actuators A*, **24**, 15-19.

7.38 Prudenziati M, Morten B and Taroni A (1981/2) Characterisation of thick-film resistor strain gauges on enamel steel. *Sensors and Actuators*, **2**, 17-27.

7.39 Erskine JC (1983) Polycrystalline silicon on metal strain gauge transducers. *IEEE Trans. Electron Devices*, **30**, 796-801.

7.40 Zwicker UT (1989) Strain sensor with commercial SAWR. *Sensors and Actuators*, **17**, 235-239.

7.41 d'Amico A and Verona E (1989) SAW sensors. *Sensors and Actuators*, **17**, 55-66.

7.42 Chetwynd DG and Smith ST (1991) High precision surface profilometry: from stylus to STM, in *From Instrumention to Nanotechnology* (eds. J.W. Gardner and H.T. Hingle), Gordon & Breach Science Publishers, Philadelphia, PA, USA, pp. 273-300.

7.43 Kittilsland G, Stemme G and Norden B (1990) A submicron particle filter in silicon. *Sensors and Actuators*, **A21-A23**, 904-907.

7.44 Martin BA, Wenzel SW and White RM (1990) Viscosity and density sensing with ultrasonic plate waves. *Sensors and Actuators*, **A21-A23**, 704-708.

7.45 Varaprasad AM and Krishnan R (1988) PZT-polymer composite for transducers of hydrophone systems. *Sensors and Actuators*, **14**, 361-368.

7.46   Royer M, Holmen JO, Warm MA, Aadland OS and Glenn M (1983) ZnO on Si integrated acoustic sensor. *Sensors and Actuators*, **4**, 357-362.

7.47   Berqvist J and Rudolf F (1990) A new condenser microphone in silicon. *Sensors and Actuators*, **A21-A23**, 123-125.

7.48   Okuyama M and Hamakawa Y (1986) Si-monolithic miniature ultrasonic sensor using $PbTiO_3$ thin film prepared on a Si or SiO_2 cantilever, in *Microsensors and Microactuators* (eds. R.T. Howe, R.S. Muller, A.P. Pisano and R.M. White), IEEE Press, USA.

7.49   Monchalin J-P (1986) Optical detection of ultrasound. *IEEE Trans. on UFFC*, **33**, 485-499.

Problems

7.1   Construct a simple lumped system model of an electrostatically-driven microflexure (i.e. a cantilever beam) with capacitive displacement sensing. Using equation (7.4) or otherwise, determine the displacement of the microstructure to a sinusoidal driving voltage.

7.2   From the model that you constructed in problem 7.1, calculate the current flowing through the sensing capacitor using typical geometric and mechanical parameters for a polysilicon resonant microflexure.

7.3   Why is single-crystal silicon not used as a piezoelectric material but is used as a piezoresistor? What materials may be used in a piezoelectric sensor? What problems may arise from their application in a real microsensor?

7.4   State three parameters, other than acceleration, that can be measured using a silicon resonant microstructure. Describe the basic principles involved together with any factors that could limit the application of such a type of microsensor. *(You may wish to refer to other chapters in order to answer this problem)*

8. Magnetic Microsensors

Objectives

- [ ] To introduce the topic of magnetic microsensors

- [ ] To review the basic types

- [ ] To examine their underlying principles

8.1 Basic Considerations and Definitions

8.1.1 Introduction to magnetism

Magnetic fields naturally occur all around us. For instance, celestial bodies generate magnetic field strengths that vary from about 2×10^{-5} A/m for a remote galaxy to about 600 A/m at the surface of the Sun or at the poles of Jupiter [8.1]. Small magnetic fields are also generated by a biological system. For example, the microcurrents that flow in our brain, heart and muscle tissues, generate small magnetic fields in the region of 10^{-2} A/m. Figure 8.1 illustrates the scale of magnetic field strengths which range from 10^{-9} A/m to about 10^{8} A/m. Thus larger magnetic fields are man-made (technical) and arise in transmission cables, power transformers (typically 10^{1} to 10^{5} A/m), soft or hard ferromagnetic material (typically 1 to 10^{6} A/m) and conventional solenoids or superconducting coils (e.g. 10^{6} to 10^{7} A/m).

The units of magnetic field strength H are amperes per metre (A/m) whereas the units of magnetic induction or magnetic flux density B are tesla (T). The magnetic flux density B is 1 tesla when a magnetic flux Φ of 1 weber passes perpendicularly through an area of 1 m². The magnetic induction B in a material is related to the magnetic field strength H by,

$$B = \mu^{B} H = \mu_{0}^{B} \mu_{r}^{B} H \qquad (8.1)$$

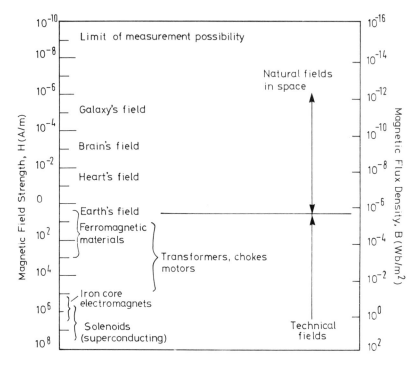

Figure 8.1 Typical levels of magnetic field strengths.

where μ^B is the magnetic permeability of the material and equals the product of the magnetic permeability of free space $\mu_0{}^B$ ($4\pi \times 10^{-7}$) and the relative permeability $\mu_r{}^B$. Both high-permeability (ferro- or ferri-) materials ($\mu_r{}^B \gg 1$), and low-permeability (dia- or para-) materials ($\mu_r{}^B \sim 1$) are used in magnetic sensors.

8.1.2 Magnetic effects for sensors

The discovery of various magnetic effects and the development of special magnetic materials has enabled the fabrication of a wide range of magnetic sensors. Table 8.1 provides a brief overview of these magnetic effects from the Joule effect discovered in 1842 to the Josephson effect discovered in 1962. These effects are exploited in various ways. In direct applications, the magnetic sensor is essentially a magnetometer. Thus it is used to measure a magnetic field strength, such as that coming from the earth, or the field modulated by magnetic material on magnetic tapes, discs, credit cards etc. In indirect applications, the magnetic sensor is used as an intermediate sensor to detect non-magnetic signals. There are many examples of indirect applications such as:

- linear and angular position sensing using Hall effect devices;
- pressure sensing using Hall effect devices;
- control of brushless DC motors;
- non-contact current sensing.

Table 8.1 Magnetic effects used in sensors.

Year	Effect	Description	Application
1842	Joule	Magnetostriction - change in shape of ferromagnet with magnetisation.	Acoustic delay line magnetometers.
1846	-	Change in Young's modulus with magnetisation.	Acoustic delay line magnetometers.
1847	Matteucci	Torsion of a ferromagnetic rod changes magnetisation.	Magnetoelastic sensors.
1856	Thomson	Change in electrical resistance with magnetic field.	Magnetoresistive sensors.
1858	Wiedemann	A torsion is created by a current-carrying ferromagnetic rod when subjected to a longitudinal field.	Torque and force measurement.
1865	Villari	Magnetisation effected by tensile or compressive strength.	Magnetoelastic sensors.
1879	Hall	Transverse voltage created across current-carrying crystal by magnetic field.	Magnetogalvanic sensors.
1903	Skin	Displacement of current to surface due to eddy current.	Position sensors (distance, proximity).
1931	Sixtus Tonks	Pulse magnetisation by large Barkhausen jumps.	Wiegand and pulse-wire sensors.
1962	Josephson	Quantum mechanical tunnelling effect between two superconducting layers.	SQUID magnetometers.

In direct applications, magnetic sensors are commonly used to detect magnetic flux densities in the range of micro- to milli-tesla which are usually man-made.

8.1.3 Classification of magnetic microsensors

The type of magnetic sensor which makes use of a large coil or ferromagnetic block is clearly not a microsensor and so will not be discussed in this chapter. Semiconductor magnetic field sensors are microsensors and they exploit the galvanometric effects due to a Lorentz force on charge carriers moving in a magnetic field (see later). Similarly, acoustic magnetic sensors exploit the magnetostrictive and magneto-elastic effects in thin or thick films. Figure 8.2 classifies magnetic microsensors in terms of their basic generating principle (e.g. acoustic) and then the type of device (e.g. magnetodiode, §8.4). The classification covers all types of semiconductor (e.g. silicon) sensors as well as those which use thin or thick active films (e.g. a SAW delay-line sensor). The Superconducting Quantum Interference Device (SQUID) is unlike other microsensors not only because of its operating principle but because of its extremely high sensitivity. A

Genus: Principles: Devices:

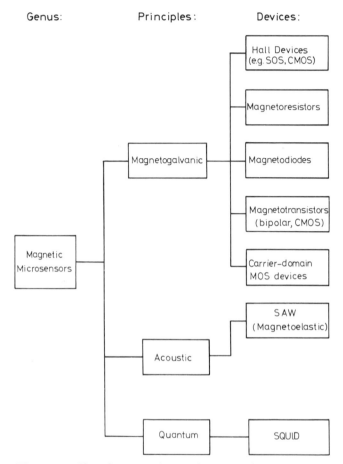

Figure 8.2 Classification scheme of magnetic microsensors.

SQUID can resolve the picotesla signals which are found in biomagnetometry and it is the most sensitive magnetic device currently available (§8.8).

8.2 Hall Devices

8.2.1 *The Hall effect*

When an electron moves inside a metal in the presence of a magnetic flux density B, a magnetic force acts upon it. This so-called Lorentz force F acts perpendicular to the electron motion (i.e. current flow) and is given (in vector notation) by,

$$F = e(v \otimes B) \tag{8.2}$$

where v is the average (drift) velocity of the electrons. The Lorentz force rotates the current flow lines through an angle θ_H known as the Hall angle. The deflection of the electrons results in positive charge forming on one side of a conducting material and a negative charge on the other. The resultant Hall electric field E_H opposes the Lorentz force and so under equilibrium conditions (in scalar notation)

$$E_{H_y} = +v_x B_z \qquad (8.3)$$

Hall-effect magnetic sensors exploit this effect. Figure 8.3 shows the general layout of a Hall plate sensor. The voltage V_H across the plate (Hall voltage) can be calculated from simple transport theory as

$$V_H = E_H w = +v_x B_z w \qquad (8.4)$$

where w is the width of the Hall plate. Relating the average drift speed v_x to the current I_x gives,

$$V_H = \left[-\frac{1}{ne} \right] \frac{I_x B_z}{d} = R_H \frac{I_x B_z}{d} \qquad \text{(ideal Hall voltage)} \qquad (8.5)$$

where n is the electron density, d the plate thickness and R_H is called the Hall coefficient. Thus the magnitude of both the Hall voltage and Hall angle are determined by the magnitude of the Hall coefficient, with

$$\tan \theta_H = \frac{E_y}{E_x} = R_H B_z \sigma \qquad (8.6)$$

where σ is the electrical conductivity of the material. In metals, such as gold or copper, the Hall coefficient R_H is small and negative ($ca.\ -1 \times 10^{-10}\ \mathrm{m^3/C}$) and close to the value predicted from this classical transport theory.

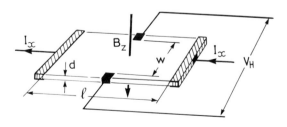

Figure 8.3 Schematic diagram of a Hall plate magnetic sensor.

The same classical transport theory can be applied to *n*-type and *p*-type semiconductors by treating the carriers as free electrons and free holes. This model is conceptually appealing and leads to Hall coefficients of

$$R_H = -\frac{1}{ne} \qquad (\text{n-type}) \tag{8.7}$$

$$R_H = +\frac{1}{pe} \qquad (\text{p-type}) \tag{8.8}$$

$$R_H = \frac{(\mu_p - \mu_n)}{n_i e(\mu_p + \mu_n)} \qquad (\text{mixture}) \tag{8.9}$$

The theoretical Hall coefficient for a semiconductor in which both electrons and holes contribute is given by equation (8.9) in terms of the electron and hole mobilities and the intrinsic carrier concentration n_i. Figure 8.4 shows the characteristic behaviour of the Hall coefficient in silicon as the doping level changes from a *p*-type to an *n*-type material [8.2].

In practice, the experimental value of the Hall coefficient differs from simple theory by a factor *r* where,

$$R_H = \frac{-r_n}{ne} \ (n\text{-type}) \ ; \ R_H = \frac{+r_p}{pe} \ (p\text{-type}) \tag{8.10}$$

r is determined from the energy-band structure and underlying scattering processes. For low doping levels at room temperature, *n*-type silicon has a value of r_n of about 1.15 while *p*-type silicon r_p has a value of about 0.7.

Finally, the Hall voltage V_H must be corrected to take account of the geometry of the plate. Using a geometry factor k_g, equation (8.5) is rewritten as

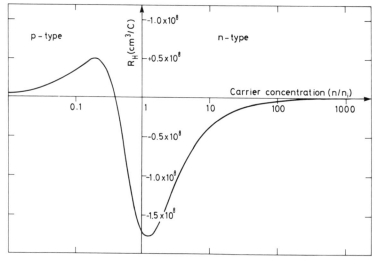

Figure 8.4 Hall coefficient in silicon when μ_p is 600 cm²/V s, μ_n is 1,500 cm²/V s and n_i is 1.6×10^{10} cm⁻³ [8.2].

$$V_H = \frac{k_g R_H I_x B_z}{d} \qquad (8.11)$$

The geometrical correction factor k_g arises because the plate is not infinitely long so the electrode contacts distort the current lines. It can be calculated from the exact geometry used [8.3].

Under strong extrinsic conditions, the Hall voltage in an n-type material is

$$V_H \approx -\frac{r_n}{ne} \cdot \frac{k_g I_x B_z}{d} \qquad (8.12)$$

Thus, Hall-effect devices are best made from a thin plate (d is small) with a low conductivity (strictly a low carrier concentration) and a careful choice of geometry so that $k_g \sim 0.7$.

The sensitivity of a Hall sensor can be defined from its supply current I_x or supply voltage V, S_I and S_V, respectively:

$$S_I \equiv \left| \frac{1}{I_x} \frac{\partial V_H}{\partial B_z} \right| = \frac{r_n k_g}{ned} \qquad (8.13)$$

$$S_V \equiv \left| \frac{1}{V} \frac{\partial V_H}{\partial B_z} \right| = \frac{S_I}{R} \qquad (8.14)$$

where R is the resistance of the device. Typical values of the sensitivities are 100 V/AT for S_I and 0.1 /T for S_V in silicon devices.

8.2.2 Integrated Hall sensors

Figure 8.5 shows the top view and cross-section of an n-type Hall plate magnetic sensor. The plate is defined by a deep p-type diffusion and this structure is fully compatible with an IC bipolar process. The current flows between the electrodes labelled 1 and 2, while the Hall voltage is measured across electrodes 3 and 4. Typical electron densities and thickness of the epitaxial layers are 10^{15} to 10^{16} cm^{-3} and 5 to 10 µm, respectively. This design produces a reasonable device with typical plate dimensions of 200 µm by 200 µm.

The resistivity of the epitaxial layer can be reduced by local ion implantation (§3.2.1). The technique can lead to a better uniformity of plate and has been used to fabricate highly linear GaAs devices [8.4].

This type of bipolar IC device measures the magnetic field vertical to the chip plane. For applications in which a magnetic field parallel to the chip plane is desired, there is the so-called "vertical" Hall device [8.5]. The structure of a vertical Hall device is shown in Figure 8.6. The structure has been used recently to make a monolithic silicon compass [8.6] using two orthogonal 100 µm square devices. The sensitivity S_I of each device is 41 V/AT with a noise level of 1×10^{-5}

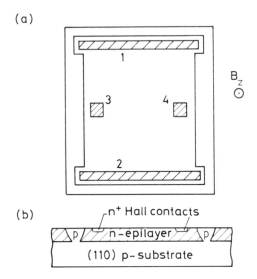

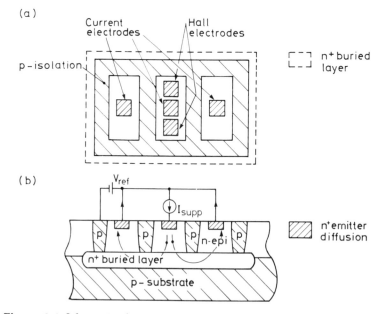

Figure 8.5 Schematic layout of a Hall plate magnetic-field microsensor.

$T/\sqrt{Hz}$ at 40 Hz. The angle of magnetic flux densities of various strengths was calculated to provide an output voltage roughly proportional to the field angle, see Figure 8.7.

Horizontal and vertical Hall magnetic-field sensors have also been fabricated

Figure 8.6 Schematic diagram of a vertical Hall plate magnetic-field microsensor [8.5] (© 1984 IEEE).

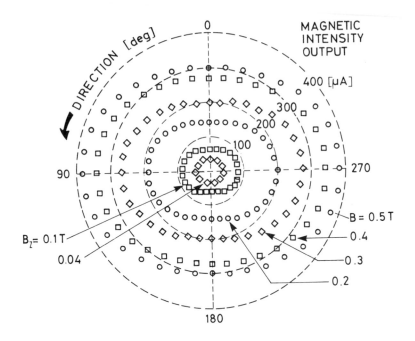

Figure 8.7 Output of a monolithic silicon magnetic compass [8.6].

using CMOS technology. For example, a 3-D vertical Hall magnetic field sensor has been realised in a standard 2 μm CMOS process [8.7]. The device can measure all three components of the magnetic flux density, (B_x, B_y, B_z), simultaneously, giving a linear response. An advantage of an MOS Hall device over a bipolar device is that its sensitivity S_I is higher at 500 to 1,000 V/AT because its ultra-thin Hall layer (~ 100 Å) more than compensates for the lower mobility.

Finally, reactive ion etching (§4.2) has been used to enhance the sensitivity of bipolar vertical Hall sensors. The resultant structure is the structure of a junction field-effect transistor, JFET. A high sensitivity S_I of about 1,250 V/AT has been reported for a vertical Hall device.

8.3 Magnetoresistors

8.3.1 Semiconductor

The resistance of a semiconductor is influenced by the application of an external magnetic field. The main effect is due to the Lorentz force which rotates the current lines by an angle θ, where

$$\theta = \tan^{-1}(\mu B_z)$$

(8.15)

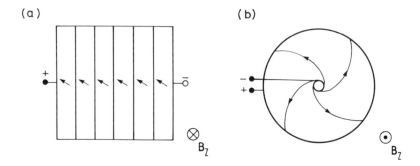

Figure 8.8 Possible configuration of a magnetoresistor: (a) parallel Hall plates and (b) a Corbino disc [8.2].

The current path thus becomes narrower and longer leading to an increase in resistance for small angles of

$$R(\theta) \approx R_0 (1 + \tan^2 \theta) \quad \Rightarrow \quad R(B_z) = R_0 (1 + k_{ar} \mu^2 B_z^2) \qquad (8.16)$$

The constant k_{ar} depends upon the aspect ratio of the plate. Ideally the plate is much wider than long. Two possible designs are a series of Hall plates or a Corbino disc as shown in Figure 8.8 (a) and (b), respectively [8.2]. Both structures short out the Hall voltage as required. The low electron mobility of silicon (~1,600 cm^2/V s) makes the effect rather small. However, InSb has an exceptionally high mobility of 70,000 cm^2/V s. Moreover, the addition of NiSb into InSb results in a structure with metallic needles that when oriented perpendicular to the (zero field) current flow act as short-circuiting plates.

Figure 8.9 shows the effect of magnetic flux density upon the resistance of commercial undoped and doped InSb/NiSb magnetoresistors. The large changes in electrical resistance result from significant Hall angles. The output from an undoped and doped InSb/NiSb magnetoresistor is temperature-dependent as shown and so an accurate measurement requires the use of calibration charts or temperature stabilisation.

8.3.2 Magnetic thin film

Electrically conductive magnetic materials exhibit the so-called anisotropic magnetoresistance effect. In these materials the magnetisation vector determines the direction along which the current normally flows. The application of an external magnetic field rotates the magnetisation vector in the film and thus the

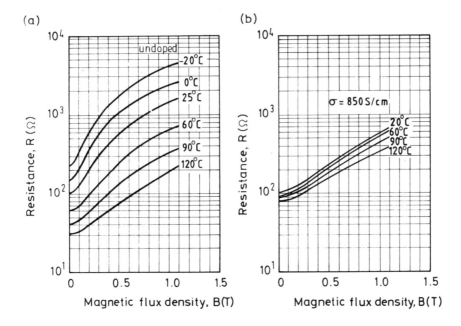

Figure 8.9 Typical performance of InSb/NiSb magnetoresistors: (a) undoped material and (b) doped material.

current path. The resistance $R(\theta)$ of the device is then related to the angle θ by the function,

$$\frac{(R(\theta)-R_0)}{(R_{n0}-R_0)} = \sin^2\theta \qquad (8.17)$$

where R_0 is the resistance of the device at zero angle and R_{n0} is a material constant. This function is rather non-linear and ideally the angle is set at 45° for a high sensitivity. This can be achieved using a barber-pole structure with, for example, a permalloy film (81% Ni, 19% Fe). The sensitivity S_V of such a device in an active bridge configuration (§2.2.1) was reported recently as 3 /T with a temperature coefficient of -3.5×10^{-3}/K [8.8].

8.4 Magnetodiodes

Figure 8.10 shows the basic arrangement of a magnetodiode. Carriers are injected into the semiconducting region of dimensions ($l \times w \times d$) from the n^+ and p^+ regions, and drift under the action of an electric field. The magnetodiode effect arises from the Suhl or magnetoconcentration effect [8.9] in which a difference in the recombination rates exists between the two interfaces, and the double injection of carriers. The injected carriers are deflected by the Lorentz force towards an edge of the region (either 1 or 2) as described previously which leads to a carrier

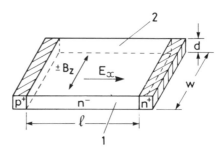

Figure 8.10 Schematic diagram of a magnetodiode.

concentration gradient perpendicular to the electrical field E_x. These effects cause a modulation of the diode current-voltage characteristic which depends upon the relative recombination rates, geometry and current load.

The difference in surface recombination rate has been generated in integrated magnetodiodes through the use of Si-Al$_2$O$_3$ (high rate) and Si-SiO$_2$ (low rate) surfaces. These so-called Silicon-On-Sapphire (SOS) magnetodiodes have the structure shown in Figure 8.11(a). For a thickness d, less than the ambipolar diffusion length and a small magnetic flux density (< 20 mT), the forward voltage sensitivity of the magnetic diode at a constant current is given by [8.10],

$$S_{V_f} = \frac{dV_f}{dB_z} = \frac{e(\mu_n + \mu_p)\tau_{eff}(\upsilon_2 - \upsilon_1)}{8kTl} \cdot V_f^2 \qquad (8.18)$$

where l is the length of the n-type silicon region, υ_1 and υ_2 are the recombination rates, μ_n and μ_p are the carrier mobilities, and τ_{eff} is the effective carrier life-time. A sensitivity of about 4.9 V/T has been reported [8.10]. SOS magnetodiodes tend to suffer from several drawbacks: first the recombination rate at the Si-Al$_2$O$_3$ surface is difficult to reproduce, secondly the sensitivity is highly non-linear, and lastly the devices are temperature-dependent. Attempts have been made to use differing mechanical roughness to determine the recombination rate difference in devices but this has also proved to be unreliable.

A slightly different principle is used in a magnetodiode-like structure that is compatible with the standard IC process such as bulk CMOS, see Figure 8.11(b). The structure looks rather like a bipolar transistor but the reverse-biased *p-n* junction (i.e. the collector) becomes the high recombination surface of the conventional SOS magnetodiode [8.11]. A magnetic field applied perpendicular to the base current deflects the current lines towards or away from the collector. Thus, as the collector current is modulated the base resistivity will change and thus the base-emitter voltage is modulated. A forward voltage sensitivity of up to 25 V/T has been reported [8.11] and thus is somewhat larger than that observed in SOS magnetodiodes.

(a)

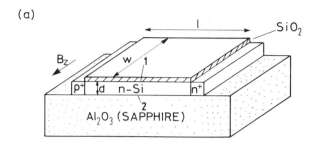

(b)

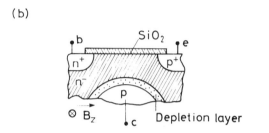

Figure 8.11 Schematic layout of two types of magnetodiodes: (a) silicon-on-sapphire, and (b) CMOS [8.11] (© 1984 IEEE).

8.5 Magnetotransistors

8.5.1 Introduction

Although silicon Hall plates can be manufactured using standard bipolar IC technology (§3.2), the output voltage can suffer from an offset due to perhaps a masking alignment error or a piezoresistive voltage caused by internal stresses. Magnetodiodes do not suffer from this offset problem but their sensitivity varies due to the poor control of the high recombination rate surface. Not surprisingly, transistor structures are also influenced by magnetic fields and so can be employed as magnetic microsensors.

The output signal of a magnetotransistor or *magnistor* is a current difference. A magnetotransistor basically consists of a current source (i.e. emitter) and several electrodes (i.e. collectors) to sink the current. It is not essential to have a split drain but it removes the quiescent signal and thus reduces the non-linearity.

There are two principles by which a magnetotransistor can operate. First, charge carriers are deflected by the Lorentz force, as in Hall plates, when the Hall field is prevented from being established. The deflection causes an imbalance in the collector currents. Secondly, the emitter injection is modulated, as in magnetodiodes, when a Hall voltage in the base modulates current injection via the emitter. The design of the magnetotransistor will determine which mechanism dominates its magnetic field sensitivity.

Magnetotransistors are usually classified according to the direction of the main current with respect to the silicon surface. When the current flows laterally, the devices are sensitive to perpendicular fields. Alternatively, when the current flows vertically, the devices are sensitive to lateral fields.

8.5.2 Lateral magnetotransistors

Figure 8.12 shows the structure of one of the first lateral magnetotransistors to be fabricated using standard bipolar technology [8.12]. The emitter current I_e is deflected by the hole Hall angle which results in a current difference across the two p^+-type collectors c_1 and c_2. Applying a positive voltage across the n^+-type base contacts b_1 and b_2, causes the base to behave as a Hall plate and a Hall field is generated. The Hall field deflects the current path by a further angle and thus contributes to the current difference. Combining these effects gives,

$$\Delta I_c = K_g (\mu_n + \mu_p) B_z I_e \qquad (8.19)$$

where μ_n and μ_p are the electron and hole mobilities, and K_g is a constant which depends upon the device geometry, bias conditions and electrical device properties.

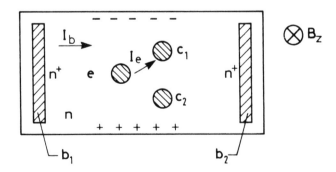

Figure 8.12 Drift-aided lateral *p-n-p* magnetotransistor [8.12].

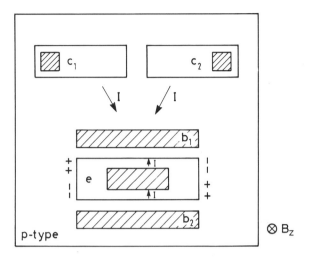

Figure 8.13   Structure of an injection-modulation lateral *p-n-p* magnetotransistor [8.13] (© 1984 IEEE).

In a dual-base dual-collector magnetotransistor the injection modulation dominates the magnetic sensitivity. As shown in Figure 8.13 the collectors are now outside the base pads [8.13]. Application of the perpendicular field sets up opposing Hall fields in the base and emitter areas. This in turn changes the injected emitter current differentially on each side and thus the currents flowing through the collectors.

Lateral magnetotransistors have also been fabricated using CMOS technology. A good discussion of Hall MOSFETs and split-drain MAGFETs may be found in [8.2]. The main advantage of employing CMOS technology can be an enhanced sensitivity.

8.5.3 Vertical magnetotransistors

In a vertical magnetotransistor, the current flows perpendicular to the silicon surface and so a sensitivity to lateral magnetic fields is achieved. There are two main types of vertical magnetotransistors: the Differential Amplification Magnetic Sensor (DAMS) and the dual-collector magnetotransistor. Figure 8.14 shows the cross-section of both structures [8.2]. The DAMS consists of two vertical *p-n-p* transistors in which the base region is shared. A <u>perpendicular</u> magnetic field generates a Hall voltage between the emitter areas which modifies the injected currents and thus the collector currents. The advantage of a DAMS is that the small Hall signal is amplified by this arrangement.

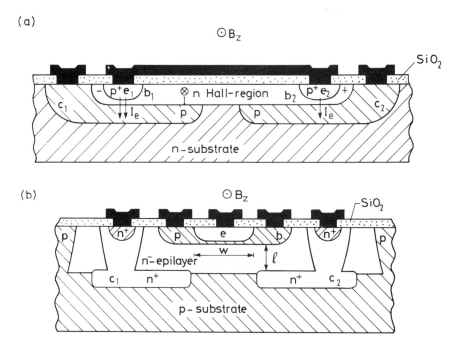

Figure 8.14 Cross-section of (a) a DAMS and (b) a dual-collector vertical magnetotransistor. From [8.2].

In a vertical dual-collector magnetotransistor, see Figure 8.14(b), the device has a current which flows perpendicular to the silicon surface and so is sensitive to lateral magnetic fields. It consists of two *p-n-p* transistors as before except that the base and emitter are now common to both devices. A lateral magnetic field deflects the electrons towards one of the n^+ collectors and thus generates a collector current difference of

$$\Delta I_c = \left[\frac{K_g l_{\text{eff}}}{w_e} \right] I_c \mu_x B_y \qquad (8.20)$$

where K_g is a constant (theoretically 2), l_{eff} is the effective deflection length (i.e. thickness of epilayer), w_e is the width of the emitter, I_c is the total collector current ($\propto I_e$) and $\mu_x B_y$ is the Hall angle.

Vertical magnetotransistors of varying designs have been reported; however, their output is quite sensitive to lithographic mask errors so Hall plate devices are generally easier to manufacture.

8.6 Carrier-domain Magnetic Field Microsensors

The carrier-domain magnetic field sensor was first proposed in 1976 as a device employing a novel type of operating principle [8.14]. It consists of a multiple *p-n-p-n* structure as shown in Figure 8.15. The central part of the device consists of a

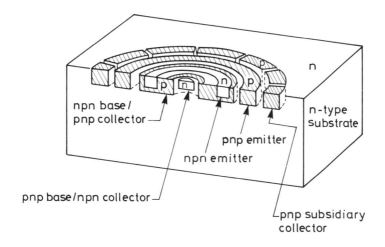

npn base /
pnp collector

n-type
substrate

pnp emitter

npn emitter

npn emitter

pnp base / npn collector

pnp subsidiary
collector

n

n-type
substrate

Figure 8.15 Rotating carrier domain magnetic field microsensor [8.14].

circular vertical p-n-p transistor and lateral n-p-n transistor. The p-type collectors that orbit the centre act as additional collectors for the lateral n-p-n transistors and detect the rotation of the current domain. The current is localised by this structure to a small domain which rotates in the presence of the magnetic field. The frequency of rotation f_r is given by

$$f_r = \frac{\mu_p B_z d}{2\pi a \tau_p} \tag{8.21}$$

where d is the radial spacing between the emitter and base regions, $\mu_p B_z$ is the Hall angle, τ_p is the electrical time constant of the base layer and a is the radius of the outer edge of the n-p-n emitter.

Unfortunately, there is a threshold magnetic field below which the device will not work and the output is very sensitive to both temperature and lithographic mask errors. However, a high current sensitivity of 3 mA/T has been reported at a driving current of 10 mA and a noise level equivalent to about 500 μT [8.15].

8.7 Acoustic Magnetic Field Microsensors

When a Surface Acoustic Wave (SAW) propagates in a thin magnetic material, a magnetoelastic coupling can occur between the magnetic spin and strain fields. Consequently, the application of an external DC magnetic field can modify the acoustic characteristics of a magnetoelastic material. The principle has been exploited in a SAW delay-line (§7.7.3) oscillator in which the change in oscillator

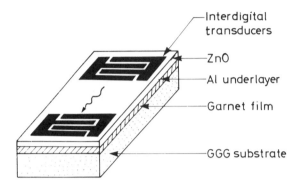

Figure 8.16 A SAW delay-line magnetic-field microsensor [8.16] (© 1987 IEEE).

frequency $\Delta f/f_0$ and acoustic phase velocity v are a function of the magnetic flux density,

$$\frac{\Delta f}{f_0} = \frac{\Delta v}{v} = F(B_z) \tag{8.22}$$

Figure 8.16 shows a SAW delay line device configured on top of a garnet film. The garnet film was epitaxially grown on a (111) Gadolinium Gallium Garnet (GGG) substrate, then, after laying down an Al sublayer, ZnO was sputtered to

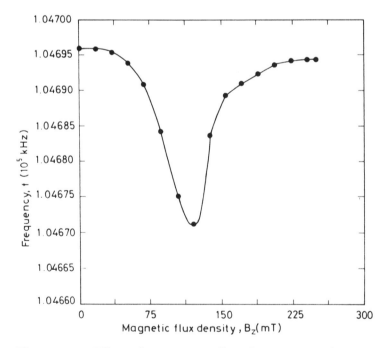

Figure 8.17   Effect of a magnetic flux density upon the resonant frequency of a SAW magnetic-field microsensor [8.16] (© 1987 IEEE).

improve the electromechanical transducer coupling [8.16]. Figure 8.17 shows the effect of magnetic flux density B_z upon the frequency of the SAW oscillator. The device has a centre frequency f_0 of 105 MHz and a wavelength of 32 µm. The magnetic field sensitivity S_H is about 1 Hz/(A/m), with a flux density resolution of 1 µT in a magnetic field of 15 A/m, i.e. about one part in 100. The implementation of such a device with digital electronics is an attractive proposition.

8.8 SQUIDs

The Superconducting Quantum Interference Device or SQUID is the most sensitive magnetometer with a magnetic inductance resolution of a few fT. Consequently, SQUIDs are able to measure extremely weak magnetic fields such as those generated by the human heart or brain ($\sim$ nT). A SQUID consists of a ring structure made of a superconducting material (e.g. $YBa_2Cu_3O_{(7-x)}$ with a critical temperature, T_c of 90 K) which is interrupted by either one (R.F. SQUID) or two (DC SQUID) Josephson junctions.

Figure 8.18 shows the basic arrangement of a DC SQUID. The superconducting ring is injected with a constant bias current I_{bias}. The current splits into two parts, I_1 and I_2, which flow through the two Josephson junctions 1 and 2. The Josephson junctions form weak links in the magnetic loop. In the absence of the weak links, the flux Φ_{ext} inside the loop is related to the integral of the external magnetic flux density B_{ext} around a surface A that encloses the loop,

$$\Phi_{ext} = \oint_{cs} B_{ext} \cdot dA \qquad (8.23)$$

The magnetic flux Φ_{in} inside the superconducting loop must be quantised in units

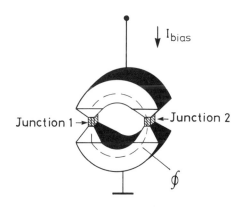

Figure 8.18 Basic configuration of a DC SQUID.

of the flux quantum Φ_0, where $\Phi_0 = h/2e$, i.e. 2.07 fWb. Consequently, the external flux is compensated internally in units of Φ_0 by a self-induced flux Φ_s, where

$$2\pi\Phi_{int} = 2\pi n\Phi_0 \text{ and } \Phi_s = LI_{cc} \tag{8.24}$$

where L is the self-inductance of the ring and I_{cc} is the circulating shielding current. Now the introduction of the two Josephson junctions results in phase differences δ_1 and δ_2 across each weak link. So the flux rule becomes,

$$2\pi n = \frac{2\pi\Phi_{int}}{\Phi_0} - \delta_1 + \delta_2 \tag{8.25}$$

We can now relate the constant bias current I_{bias} flowing through the SQUID to the sum of the currents flowing through each Josephson junction and hence the phase differences,

$$I_{bias} = (I_1 + I_2) = (I_{01}\sin\delta_1 + I_{02}\sin\delta_2) \tag{8.26}$$

Using equations (8.24) and (8.25) allows the internal flux to be related to the external flux by,

$$\Phi_{int} = \Phi_{ext} + \frac{L}{2}(I_{02}\sin\delta_2 - I_{01}\sin\delta_1) \tag{8.27}$$

Figure 8.19 shows the relationships between both the internal and external

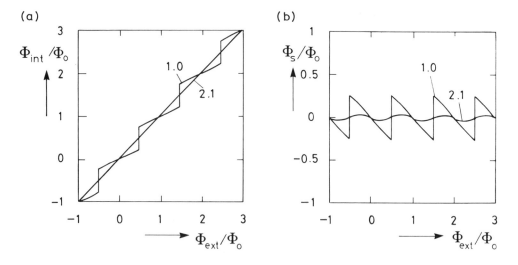

Figure 8.19 (a) Internal and (b) self-induced fluxes in a DC SQUID as a function of external flux [8.17].

fluxes and the self-induced and external fluxes (in terms of flux quanta Φ_0) [8.17]. The value of the bias current I_{bias}, relative to the base junction current I_0, is shown against the two lines plotted.

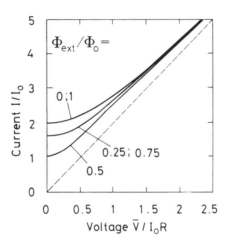

Figure 8.20 Typical current-voltage characteristic of a DC SQUID [8.17].

The time-dependent voltage across a Josephson junction $V(t)$ is given by

$$V(t) = \frac{\hbar}{(2e)} \frac{\partial \delta}{\partial t} \tag{8.28}$$

Thus, the current-voltage characteristic of a DC SQUID is usually plotted in terms of the mean voltage $\overline{V}$ relative to the peak junction current ($I_{01}=I_{02}=I_0$). Figure 8.20 shows the typical DC SQUID characteristic for relative external flux values of {0, ¼ Φ_0, ½Φ_0, ¾Φ_0 and Φ_0} [8.17]. The flux quantisation rule, equation (8.25), means that the characteristic behaviour is periodic in $n\Phi_0$.

An integrated version of the DC SQUID magnetometer has been realised. The basic arrangement is shown in Figure 8.21. The design incorporates a thin film "SQUID" ring which is shaped like a square washer. This acts as a ground plane to the input coil (usually the secondary coil of a flux transformer) thereby improving the inductive coupling between the spiral input coil and the SQUID ring. The input coil may have an inductance of 100 nH compared to a value of only 100 pH for the SQUID loop.

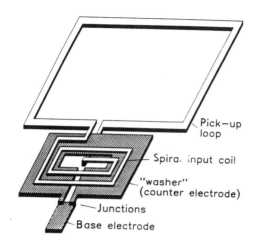

Figure 8.21 Basic configuration of an integrated SQUID magnetometer [8.17].

The SQUID read-out scheme can be quite complex. For example, the SQUID can be driven in an AC bridge (~500 kHz) with the balance condition modified by a bias modulator to give a good sensitivity to small variations.

The ultimate sensitivity of the SQUID is determined in practice by white noise or $1/f$ noise. However, its sensitivity still far exceeds all other types of magnetometer.

8.9 Relative Performance of Magnetic Microsensors

A variety of magnetic microsensors have been discussed in this chapter. These can be summarised as:

- Hall plate;
- magnetoresistors;
- magnetodiodes;
- magnetotransistors;
- carrier-domain devices;
- SAW devices;
- SQUIDs.

A useful discussion of integrated silicon magnetic-field sensors can be found elsewhere [8.18]; however, this does not include SAW devices or SQUIDs. Although it is difficult to compare all these devices, it is possible to identify some key features that will relate to their application. Table 8.2 summarises some of the key features of magnetic microsensors.

Table 8.2 Typical characteristics of magnetic microsensors.

Sensor type	Range (T)	Sensitivity	Cost	Comments
Hall effect	10^{-3}-10^{-9}	5-10 (%/T)	Low	Good linearity. Si is much less sensitive than InSb. Small offsets < 10 mT.
(Commercial)	0-1	200 (V/T)	Low	Siemens KSY10. ±0.7% linearity.
Magnetoresistor	10^{-1}-10^{-10}	~500 (%/T)	Low	Less linear but more sensitive than Hall ICs.
(Commercial)	0-1	700 (%/T)	Low	Siemens F830L 100E. Non-linear (parabolic) and higher T_c.
Magnetodiodes	$>10^{-3}$	5 (V/T) 25 (V/T)	Low	SOS device. CMOS device.
Magneto-transistors	$>10^{-3}$	0.5-2 (%/T) 7 (%/T) 2-5 (%/T)	Low	Lateral drift aided p-n-p. Lateral injection p-n-p. Vertical deflection n-p-n. Difficult to make. Temperature sensitive.
Carrier-domain	-	~100 (kHz/T)	Medium	Lateral circular n-p-n. Currently research level only.
SQUID	10^{-6}-10^{-14}	~Φ_0	High	Extremely sensitive but need superconducting temperatures.

Hall plates, magnetoresistors and SQUIDs are commercially available micromagnetometers in competition with larger devices such as flux gate and inductive coils.

There are numerous applications of magnetic-field sensors and the device sensitivity generally determines these. For example, the very high sensitivities of SQUIDs permit various applications in biomagnetism such as the measurement of the magnetic fields generated by our brain and heart and in aerial surveys for mineral oil/gas exploration.

Table 8.3 shows some of the applications of magnetic-field microsensors [8.17]. Flux gate and inductive coil magnetometers are added for the sake of completeness even though they are not microsensors. Other types of important conventional magnetic sensors are exploited in low-cost mechanical sensors, such as in linear displacement sensors (§7.3). The displacement ranges are small but find many applications in engineering. For example, the typical range of a conventional Hall effect sensor is 1 cm to 10 μm and has a good linearity. The performance of these devices makes them competitive with other conventional inductive (e.g. LVDT) displacement sensors as well as magnetic-field microsensors.

Table 8.3 Some applications of magnetic-field sensors.

Area	Hall	Magnetoresistive	SQUID	Flux gate/ inductive coil[1]
Biomagnetism (e.g. heart/brain signals)	x	x	✓✓	✓
Geomagnetism (e.g. Earth's fields)	x	✓	✓	✓✓
Identification (e.g. card reading)	✓✓	✓✓	x	✓✓
Laboratory measurements	✓✓	x	x	✓
Non-destructive testing	x	✓	✓	✓✓
Submarine and naval communication	✓	x	✓✓	✓✓

Suggested further reading

Readers are referred to the following texts that provide some background information for this chapter on magnetic microsensors:

Bleaney BI and Bleaney B: *Electricity and Magnetism* (1976). 3rd edition published by Oxford University Press, UK. ISBN 0-198511418. (762 pages. Fundamental introduction to magnetic phenomena in materials)

Göpel W, Hesse J and Zemal JN: *Sensors: A Comprehensive Survey, Vol. 5: Magnetic Sensors* (1989). Published by VCH Publishers Inc., Germany. ISBN 3-527267719. (513 pages. Comprehensive overview of all types of magnetic sensors)

References

8.1   Boll R and Overshott KJ (1989) *Sensors: A Comprehensive Survey, Vol. 5: Magnetic Sensors* (eds. W. Göpel, J. Hesse and J.N. Zemel), VCH Publishers Inc., Germany, 513 pp.

8.2   Middelhoek S and Audet SA (1989) *Silicon Sensors*, Academic Press, London, Ch. 5, p. 209.

8.3   Popovic RS (1989) Hall-effect devices. *Sensors and Actuators*, **17**, 39-53.

8.4   Hara T (1982) Highly linear GaAs Hall devices fabricated by ion implantation. *IEEE Trans. on Electron Devices*, **29**, 78-82.

8.5   Popovic RS (1984) The vertical Hall-effect device. *IEEE Electron Device Letters*, **5**, 357-358.

8.6   Maenaka K, Tsukahara M and Nakamura T (1990) Monolithic silicon magnetic compass. *Sensors and Actuators A*, **21-23**, 747-750.

8.7   Paranjape M and Filanovsky I (1992) A 3-D vertical Hall magnetic-field sensor in CMOS technology. *Sensors and Actuators A*, **34**, 9-14.

---

[1] Not a microsensor but an important conventional field detector.

8.8   Rottman F and Dettman F (1991) New magnetoresistive sensors: engineering and applications. *Sensors and Actuators A*, **25-27**, 763-766.

8.9   Suhl H and Shockley W (1949) Concentrating holes and electrons by magnetic fields. *Phys. Rev.*, **75**, 1617-1678.

8.10 Lutes OS, Nussbaum PS and Aadland OS (1980) Sensitivity limits in SOS magnetodiodes. *IEEE Trans. on Electron Devices*, **27**, 2156-2157.

8.11 Popovic RS, Baltes HP and Rudolf F (1984) An integrated silicon magnetic field sensor using the magnetodiode principle. *IEEE Trans. on Electron Devices*, **31**, 286-291.

8.12 Davies LW and Wells MS (1970) Magneto-transistor incorporated in a bipolar IC. *Proc. ICMCST, Sydney, Australia*, pp. 34-35.

8.13 Vinal AW and Magnari NA (1984) Operating principles of bipolar transistor magnetic sensors. *IEEE Trans. on Electron Devices*, **31**, 1486-1494.

8.14 Gilbert B (1976) Novel magnetic field sensor using carrier domain rotation: proposed device design. *Electron Lett.*, **12**, 608-610.

8.15 Goicollea JI, Muller RS and Smith JE (1984) Highly sensitive silicon carrier-domain magnetometer. *Sensors and Actuators*, **5**, 147-167.

8.16 Hanna SM (1987) Magnetic field sensors based on SAW propagation in magnetic films. *IEEE Trans. on UFFC*, **34**, 191-194.

8.17 Koch H (1989) SQUID sensors, in *Sensors: A Comprehensive Survey, Vol. 5: Magnetic Sensors* (eds. W. Göpel, J. Hesse and J.N. Zemel), pp. 381-445.

8.18 Kordic S (1986) Integrated silicon magnetic-field sensors. *Sensors and Actuators*, **10**, 347-378.

Problems

8.1   Estimate the Hall voltage and Hall angle generated in an *n*-type silicon Hall plate device by a magnetic flux density of 0.1 T. The plate has a width of 100 µm and depth of 10 µm. State any assumptions that you make.

8.2   What are the general merits of using a magnetodiode or a magnetotransistor rather than a Hall plate device to measure a magnetic field? Why do you think this potential advantage has not been fully realised?

8.3   State the type of magnetic microsensor that you would choose to measure: (a) the magnetic field around the human heart and (b) the field leakage around a powerful electromagnet. In each case explain the reasons for your choice.

8.4   It is not uncommon to use a magnetic sensor to measure indirectly a non-magnetic property. Discuss three possible examples. *(You may wish to refer to other chapters in order to answer this problem)*

9. (Bio)chemical Microsensors

Objectives

- [ ] To introduce the subject of (bio)chemical microsensors

- [ ] To present their basic principles

- [ ] To review their performance

9.1 Basic Considerations and Definitions

9.1.1 Definitions

A (bio)chemical sensor is a device which is capable of converting a chemical (or biological) quantity into an electrical signal. Chemical species are usually present in mixtures where the sample matrix might be gas, liquid or solid. For example, a chemical sensor may detect the presence of hydrogen in air (i.e. a gas sensor), the presence of water vapour in air (i.e. a humidity sensor) or the presence of ions in water (e.g. a potassium ISFET). A chemical sensor might also detect the presence of more complex molecules such as sugars or proteins in a liquid. A chemical sensor which detects a biological quantity is normally referred to as a biochemical sensor or biosensor.

9.1.2 Classification of (bio)chemical sensors

The different types of (bio)chemical sensors may be classified according to the sensing principle by which they detect the chemical measurand. For example, the interaction of a chemical species (X) with the sensing material (M) may often be described by the following reaction,

$$X + M \underset{k_b}{\overset{k_f}{\Longleftrightarrow}} (X.M) \tag{9.1}$$

where k_f and k_b are the forward and backward reaction-rate constants, respectively. When the reaction liberates (or abstracts) heat due to a change in the enthalpy, this can be detected calorimetrically. Alternatively, when the reaction is accompanied by a liberation (or abstraction) of charge, this can be detected conductimetrically, potentiometrically or amperometrically.

The most common principles exploited in (bio)chemical sensors are listed in Table 9.1 and can be used to classify a chemical sensor (§1.2). The table also gives the measurand and the type of sensor which exemplifies each class.

Table 9.1 Principles, measurands and typical examples of (bio)chemical sensors.

Principle	Measurand	Typical sensor
Conductometric	Resistance/conductance	Tin oxide gas sensor
Potentiometric	Voltage/e.m.f.	Ion selective FET for pH
Capacitive	Capacitance/charge	Polymeric humidity sensor
Amperometric	Current	Electrochemical cell
Calorimetric	Heat/temperature	Pellistor gas sensor
Gravimetric	Mass	Piezoelectric or SAW sensors
Optical	Path length/absorption	Infra-red detector for methane gas
Resonant	Frequency	Surface plasmon
Fluorescent	Intensity	Fibre-optic

The principles can now be used to classify all (bio)chemical microsensors according to their device type as shown in Figure 9.1. All of these microsensors are modulating rather than self-generating in nature (§1.2). There is a type of chemical sensor that is self-generating, namely, the electrochemical cell which generates an electrochemical potential that depends on the gas concentration. However, this type of chemical sensor has yet to be miniaturised and so is not included in this discussion.

The selectivity of a chemical sensor depends upon the nature of the reaction mechanism. Ideally, a chemical sensor only responds to one chemical species despite the presence of many others. Selectivity may then be obtained through the choice of a suitable active material (M) which reacts preferentially with the measurand (X). Thin or thick layers of these active materials can be utilised in a variety of sensor types (§4.5.2). Table 9.2 gives some of the inorganic and organic materials that are commonly used in (bio)chemical microsensors. The sensor types are discussed in detail in later sections.

Perhaps the best method of obtaining a chemical selectivity is to use molecular shape recognition. This strategy is commonly used in biological sensors or biosensors to sense specific molecular structures. The principle of using the molecular shape is sometimes referred to as a key-lock system [9.1]. Figure 9.2 shows the general arrangement of a biosensor in which biological molecules are converted into an electrical signal. The filter is a material which helps remove any undesirable chemical species so that the active material can provide better selectivity.

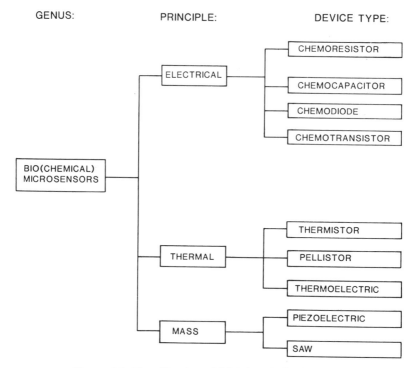

Figure 9.1 Classification of (bio)chemical microsensors.

The selectivity may also be enhanced through the use of enzymes to promote specific biological reactions, or catalysts to promote specific chemical reactions. For instance, the metal oxide semiconducting materials given in Table 9.2 are

Table 9.2 Some common materials used in (bio)chemical microsensors.

Active material	Examples	Sensing principles	Measurands
Thin oxide layer	SnO_2, ZnO	Surface conductance	Combustible gases
Thick porous oxide layers	SnO_2, ZnO, TiO_2	Bulk conductance	Combustible gases
Catalytic metals	Pd-TiO_2, Pd-MOS, Pd-MOSFET	Surface potential Threshold voltage	H_2, CO H_2, NH_3, CO
Ion-selective devices	AgCl, AgBr	Electrochemical potential	Cl^-, Br^-
Catalytic coating	ThO_2/Al_2O_3	Heat of combustion	H_2, CH_4
Organic films	Substituted phthalocyanines	Bulk conductance	NO_x
Langmuir-Blodgett films	Steric acid	Piezoelectric/SAW	Various polar molecules
Conducting polymers	Poly(pyrrole)	Bulk conductance /mass	Polar compounds, NH_3

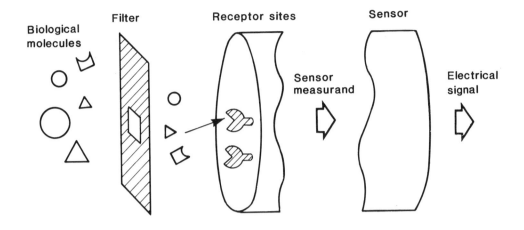

Figure 9.2 Principle of a biosensor. Adapted from [9.1].

typically doped with a few percent of reactive metals such as Pd, Pt or Rh to enhance their selectivity.

Selectivity is a desirable feature of a chemical sensor but may be difficult to achieve in practice due to the low specificity of known chemical reactions.

9.2 Chemoresistors

9.2.1 Introduction

Perhaps the simplest type of chemical sensor is the chemoresistor in which a change in the electrical conductivity of a chemically-sensitive layer is measured. Figure 9.3 shows the schematic arrangement of a chemoresistor. The inert

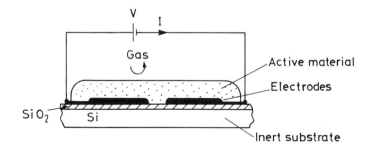

Figure 9.3 General structure of a chemoresistive gas sensor.

substrate can be made of, for example, alumina or SiO_2 and upon it a pair of electrodes is deposited. (An interdigitated electrode structure may be employed for low conductivity films.) A variety of active materials have been exploited within chemoresistors but the most common ones are metal oxides, organic crystals and more recently conducting polymers. Both the geometry and microstructure of the film influence the performance of this type of sensor as studies made on thin single-crystal films through to thick porous polycrystalline films have demonstrated.

9.2.2 Metal oxide gas sensors

In 1953 Brattain and Bardeen discovered that the adsorption of a gas onto the surface of a semiconducting material can produce a large change in its electrical resistivity. This effect has since been observed in a host of metal oxide materials such as ZnO, TiO_2 and In_2O_3. However, the most commonly used oxide material in gas sensors to date has been SnO_2. Non-stoichiometric tin oxide under certain conditions behaves as an n-type semiconductor in an environment containing oxygen (e.g. air). The principle is based upon the initial reversible reaction of atmospheric oxygen with lattice vacancies in the tin oxide and the concurrent reduction in electron concentration n'. This reaction produces various oxygen species (denoted by the constant m) at different temperatures and oxygen pressures, i.e. O_2^-, O^-, O^{2-}, which can then react irreversibly with certain combustible species (X). For example,

$$\tfrac{m}{2}O_2 + [\text{vacant site}] + e^- \xrightarrow{\ k_1\ } \left(O_m^-\right)_{site} \tag{9.2}$$

$$X + \left(O_m^-\right)_{site} \xrightarrow{\ k_2\ } \left(XO_m\right)_{site} + e^- \tag{9.3}$$

For example, the measurand X could be CH_4 which reacts with the lattice oxygen to produce CO_2 and water vapour with an increase in the carrier concentration of n''. The simple theory predicts an increase in the electrical conductivity $\Delta\sigma$ of the material which can be related to the increase in carrier concentration n'' and, from the reaction kinetics, to a fractional power r of the measurand concentration [X],

$$\Delta\sigma = \mu_n e n'' \ \propto \ [X]^r; \quad \text{where } 0.5 < r < 1 \tag{9.4}$$

Although the modulation in carrier concentration can strongly affect the surface conductance in thin single-crystal films, it is unable to explain the large effect observed in thick metal oxide layers. Instead the model of conduction must be extended to consider a concentration-dependent mobility term which arises from polycrystalline oxides having an open granular structure into which the gas can diffuse and react at the surfaces of discrete granules, see Figure 9.4. The electron mobility μ_n is then governed by the potential barrier between neighbouring grains in a hopping model. The conductance G between grains now depends upon the height of the potential barrier ϕ_B, where in the simplest model,

$$G = G_0 \exp\left(-\frac{\phi_B}{kT}\right) \qquad (9.5)$$

where G_0 is a weakly temperature-dependent constant. The modulation of the barrier height $\Delta\phi_B$ in a gas can be related to the initial and final carrier concentrations given in equations (9.2) and (9.3) by,

$$\Delta\phi_B \sim \ln\left(1 + \frac{n''}{n'}\right) \qquad (9.6)$$

where it is assumed that the bulk conductance is dominated by the gas-sensitive lattice sites. Finally, the carrier concentrations can be related back via the reaction-rates in equations (9.2) and (9.3) to the gas concentration, whence

$$\Delta G \propto \frac{k_2}{k_1}[X]^r \qquad (9.7)$$

where k_1 and k_2 are the reaction-rates in equations (9.2) and (9.3), respectively. The sensitivity and selectivity of the chemical sensor is now governed by the reaction ratio (k_2/k_1).

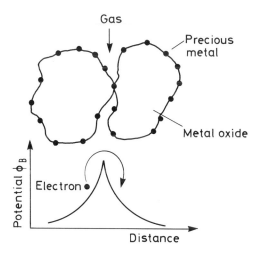

Figure 9.4 Charge transfer mechanism in a semiconducting oxide chemoresistor. The top figure shows the presence of a catalytic metal on the surface of two oxide grains. The bottom figure shows the potential barrier which forms between the grains and must be thermally "hopped" over.

As stated earlier, it is common to add a low percentage of a catalytic precious metal (e.g. Pt, Pd). This tends to enhance the modulation amplitude of the intergranular potential barrier and help promote the specificity.

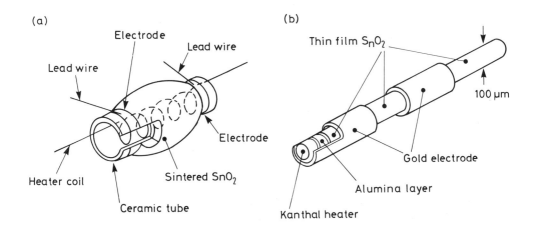

Figure 9.5 Commerical tin oxide gas sensors: (a) series-8 thick film device and (b) series-5 microdevice.

Tin oxide gas sensors have been manufactured and sold by Figaro Engineering Inc. (Japan) since the early 1970s. The most successful 8-series sensor consists of a heater coil inside an alumina ceramic tube which is coated with a thick sintered film of tin oxide, see Figure 9.5(a). Table 9.3 shows the range of gas sensors available from Figaro Inc. together with their target measurand and detection range.

Table 9.3 Commercial tin oxide gas sensors [9.3].

Model	Category	Measurand	Range (ppm)
TGS 815	Combustible gases	Methane	500 to 10,000
TGS 821	Combustible gases	Hydrogen	50 to 1,000
TGS 824	Toxic gases	Ammonia	30 to 300
TGS 825	Toxic gases	Hydrogen sulphide	5 to 100
TGS 822	Organic solvents	Alcohol, toluene	50 to 500
TGS 830	CFCs	R-113, R-22	100 to 3,000
TGS 800	Air quality	Cigarette smoke	< 10
TGS 550	Odour	Sulphur compounds	0.1 to 10

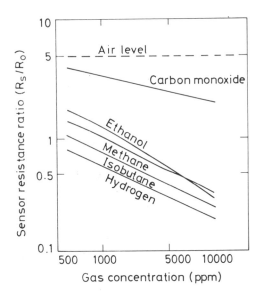

Figure 9.6 Typical response of a sintered tin oxide (commercial) gas sensor to combustible gases relative to its response to 1,000 ppm of methane in air [9.3].

Figure 9.6 shows the typical and similar response of a TGS 815 gas sensor to methane and various other gases. The high power consumption of these sensors (~1 W) is undesirable and micro-versions have been made which include the Figaro 5-series sensor with only a 75 mW power consumption, see Figure 9.5(b).

Micromachined tin oxide chemoresistors have been made in several laboratories, for example Figure 9.7 shows a photograph of a low-power thin film (100 nm) tin oxide gas sensor made at the University of Warwick, UK [9.4]. However, the lack of specificity and poor stability of tin oxide microsensors has limited their commercial applications so far to gas alarms rather than gas analysis.

Figure 9.7 Photograph of a thin film polycrystalline tin oxide gas microsensor.

9.2.3 Organic gas sensors

The use of an organic rather than an inorganic active material (e.g. SnO_2) in a gas microsensor is an attractive possibility. First of all, organic materials can be readily chemically modified to obtain a higher degree of specificity. Secondly, there is some evidence that organic materials have a higher sensitivity to pollutant and reactive gases (ppb to ppm). Thirdly, the processability of organic materials is often superior to that of inorganic oxides (which tend to have high melting points) and thus they can be deposited in a simpler manner (§4.5). Finally, organic materials operate below 150°C whereas inorganic materials usually operate at high temperature, e.g. ~ 400°C for SnO_2.

Several classes of organic materials have been studied for application in chemoresistive microsensors. The main two classes are organic crystals and conducting polymer films. Phthalocyanines form an important group of organic crystals which are thermally stable at temperatures of up to about 500°C and essentially behave like a metal oxide semiconductor [9.5]. Figure 9.8(a) shows the molecular structure of metal-free phthalocyanine (H_2Pc). The central hydrogen atoms can be replaced by metallic ions, such as Pb^{2+}, Cu^{2+}, Mg^{2+}, to improve the response to specific gases. Thin phthalocyanine films can be deposited using vacuum sublimation, although thick films can be screen-printed (§4.5.5) and ultra-thin films deposited using a Langmuir-Blodgett (§4.5.3) technique [9.6, 9.7]. PbPc has been extensively studied and shows a high sensitivity to electron-donating

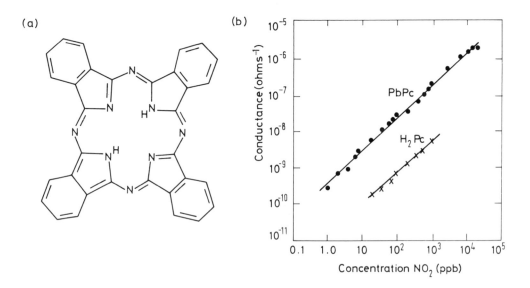

Figure 9.8 (a) Structure of metal-free phthalocyanine, and (b) typical response of metal Pc films to NO_2.

Pyrrole   Thiophene   Indole   Furan

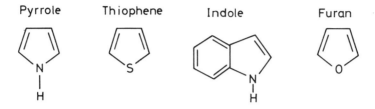

Figure 9.9 Chemical structures of some common monomers.

gases such as NO_2 [9.8]. Figure 9.8(b) shows the typical response of vacuum-sublimated PbPc and H_2Pc films to NO_2 at 150°C. The response time and recovery time of Pc films are strongly temperature-dependent, being about 100 s and 150 s at 170°C, respectively. The main problem with Pc gas sensors is that the recovery time is slow due to the presence of strongly adsorbing sites. However, specialised operating methods, such as temperature cycling, may overcome this inherent difficulty.

Conducting polymers have attracted considerable interest since their discovery in the early 1980s. Initial studies concentrated upon their application in electronic devices such as diodes, transistors, electrochromic displays and batteries. These materials have attracted widespread interest due to their range of film properties and ease of processing [9.9]. More recently, studies have been investigating the use of conducting polymers as thin film chemoresistors. Heteroaromatic monomers (or their derivatives) such as pyrroles, thiophenes, indoles and furans (see Figure 9.9) undergo electrochemical oxidation to yield conducting polymer films (§4.5.4). These films can be deposited across interdigitated electrodes to produce gas-sensitive chemoresistors [9.10]. Figure

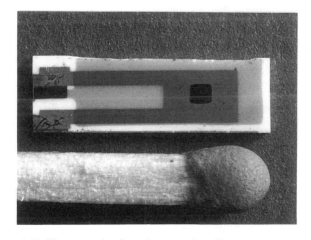

Figure 9.10 Photograph of a polymeric thin film gas microsensor.

9.10 shows a photograph of a poly(pyrrole) film grown electrochemically across a 10 µm gap between gold electrodes on a silicon substrate.

Conducting polymers are sensitive to a range of polar molecules at room temperature [9.11] and more recent reports suggest that a sensitivity down to 0.1 ppm is possible. This makes conducting polymers a potentially important active material for applications in low-power gas sensing and in odour-sensing electronic noses (§12.5.3).

9.3 Chemocapacitors

It is also possible to measure the gas-sensitivity of the dielectric constant of a film using the electrode arrangement shown in Figure 9.3. For example, a spin-coated polyphenylacetylene conducting polymer film responds to a variety of gases such as CO, CO_2, N_2 and CH_4 [9.12]. Capacitance changes in chemocapacitors are typically in the range of pF and depend strongly upon the operating frequency, and ambient conditions such as temperature and humidity. There are now several commercial hygrometers that employ a thin polymer coating in a capacitor.

Figure 9.11 shows the structure and response of a commercial polymeric humidity microsensor and its typical response. The Minicap 2 (Panametrics, USA) humidity sensor is a robust, low-cost microsensor mounted on a TO-18 header. Its characteristic performance is summarised below in Table 9.4. The interface circuit is based upon a 7556 CMOS timer circuit in which the polymer sensor changes the RC time-constant and hence the pulse-width. The output signal is then converted to a DC voltage which is proportional to the Relative Humidity (R.H.).

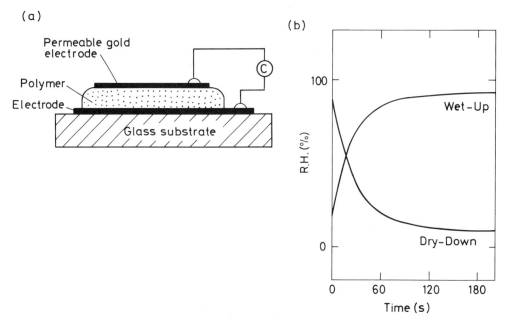

Figure 9.11 (a) Structure, and (b) typical response of a commercial humidity capacitive microsensor.

Table 9.4 Characteristic performance of a commercial polymer humidity sensor.

Range (% R.H.)	Capacitance[1] (pF)	Operating temp. (°C)	Response time, τ_{90} (s)	Linearity (%)	Stability %/yr
5 to 95	207	−40 to +180	60	±1	±1

9.4 Chemodiodes

9.4.1 Inorganic Schottky devices

The gas-sensitivity of the electrical characteristics of *p-n* diodes and Schottky junctions has also been investigated. By far the most effort has been directed towards the behaviour of inorganic and organic Schottky diodes. The work function of some catalytic metals (§4.5), such as palladium and platinum, is modified by the presence of chemical species at its surface. Thus, a chemically-sensitive Schottky diode comprises a metallic electrode (e.g. Pt) and an oxide (e.g. TiO_2, ZnO) counter electrode. A chemodiode is often run at an elevated temperature so that, for example, hydrogen can dissociate to form a dipole layer which in turn shifts the height of the junction barrier. The characteristics of these diodes are subject to considerable variability due to the nature of the metal oxide layer. Some success has been achieved through the fabrication of a tunnelling diode where a thin oxide layer (~2 nm) exists between the catalytic metal gate and a semiconducting silicon layer. A thicker oxide layer will negate the tunnelling current and so it is necessary to measure the *C-V* rather than *I-V* diode characteristics of thick oxide chemodiodes. Although Pd/*p*-SiMOS diodes have been reported as sensitive to hydrogen, most of the research effort has been directed towards the development of MOSFET structures for gas sensors (§9.5.1) and ISFET structures for detecting ionic species in liquids (§9.5.2).

9.4.2 Organic Schottky devices

The application of inorganic chemodiodes has been somewhat limited due to a poor specificity and poor stability. Consequently, scientific attention has focused upon the use of organic diodes as an alternative. The work function of conducting polymers such as poly(pyrrole) is known to change in the presence of organic vapours [9.13]. This effect has been exploited in Schottky diodes where in general a low work function metal (e.g. Cu) forms a rectifying contact with a *p*-type polymeric semiconductor (e.g. polypyrrole). The stability of polymer Schottky diodes is often questionable [9.14]; however, some success has been reported recently using a planar microstructure. Figure 9.12(a) shows the structure of an Al/poly(3-octylthiophene) device [9.15]. The diode characteristics were found to be sensitive to various gases including NH_3 and NO_x at the 100 ppm level. Figure 9.12(b) shows its *I-V* characteristics for 100 ppm of NH_3 and NO_x. Substantial differences can be seen in the forward and reverse currents for the different gases which would provide a superior specificity to most inorganic diodes.

---

[1] At 25°C, 33% R.H., and 100 kHz.

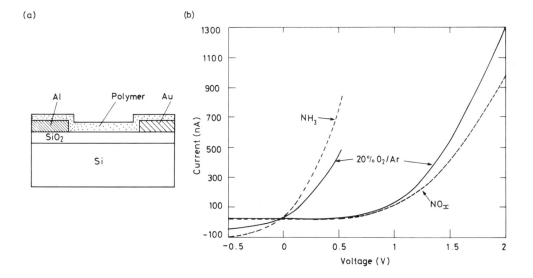

Figure 9.12 (a) Structure, and (b) response of an organic chemodiode to 50 ppm of NO_x and 100 ppm of NH_3 [9.15].

9.5 Chemotransistors

9.5.1 MOSFET gas sensors

Figure 9.13 shows the structure of a MOSFET in which the gate is made of a gas-sensitive metal, e.g. palladium. These microsensors, first proposed by Lundström in 1975 [9.16], exhibit a threshold voltage shift ΔV_T which depends upon the gas concentration according to,

$$\Delta V_T = \frac{\Delta V_{max}}{1 + k/\sqrt{[H_2]}} \tag{9.8}$$

These devices are particularly sensitive to hydrogen down to the ppm level with a maximum shift in threshold voltage ΔV_{max} of about 0.5 V. Figure 9.14 shows the typical response of a Pd-gate MOSFET to hydrogen.

The use of other gate materials (e.g. Ir and Pt) and operating the devices at different temperatures has led to reasonable specificity to gases such as NH_3, H_2S and ethanol. However, the commercial success of the MOSFET chemotransistor has been limited once again through their poor stability when compared to other types of chemical sensors. This is not true for the gateless silicon FET or Ion Selective FET (ISFET) that was first proposed by Bergveld in 1970 [9.17].

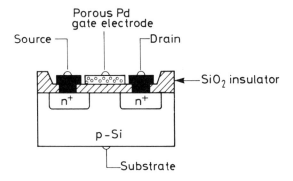

Figure 9.13 Structure of a Pd gate gas-sensitive MOSFET.

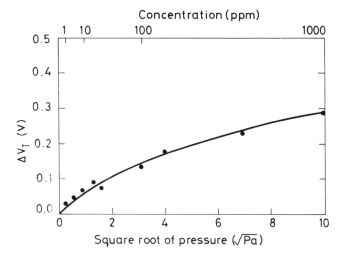

Figure 9.14 Response of MOSFET gas microsensor [9.16].

9.5.2 ISFETs

An ISFET is structurally like a FET in which the gate electrode is separated from its substrate by an electrolyte. Figure 9.15 shows the basic configuration of an ISFET. The gate electrode becomes the reference electrode in an electrochemical cell. An SiO_2 layer is often covered with an ion-selective membrane to enhance the selectivity of the ISFET. In its simplest form the ISFET is operated as a pH sensor,

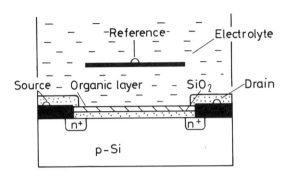

Figure 9.15 Schematic diagram of an ISFET.

where the threshold voltage shift ΔV_T depends upon the concentration of hydrogen ions in the electrolyte [9.18]. Although SiO_2 can be used as the gate oxide, a more linear response is achieved from Al_2O_3 as shown by Figure 9.16.

The pH of a solution (i.e. the effective concentration of hydrogen ions) determines the electrochemical potential of the gate (reference) electrode and thus the potential shift ΔV_T. The dashed line in Figure 9.16 shows the ideal ISFET response,

$$\Delta V_T = k_1 + k_2\left[H^+\right] = k_3 + k_4 \cdot pH \tag{9.9}$$

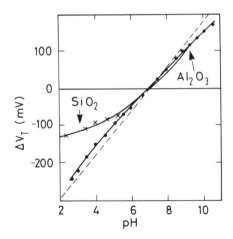

Figure 9.16 pH sensitivity of SiO_2 and Al_2O_3 ISFETs [8.2].

where the constants, k_1 to k_4, are related to the temperature of the solution and the chemical composition of the buffer material. Clearly, the alumina gate material gives a near ideal response.

The development of different types of membranes has led to the application of ISFETs to detect other cations like Na^+ and K^+. Moreover, the use of biological materials such as enzymes has resulted in biosensors. Commercial success has now been realised in ISFETs which detect pH and there is some promise in detecting substances like glucose, urea and cholesterol [9.18].

9.6 Thermal Chemical Microsensors

9.6.1 Thermistors

Thermal chemical sensors detect the heat released or absorbed, ΔE_h, when a reaction takes place. This change in enthalpy causes a change in temperature ΔT which can be monitored. In the ideal case of a thermally isolated (adiabatic) system, the change in temperature is given by

$$\Delta T = \frac{-\Delta E_h}{C_p} \tag{9.10}$$

where C_p is the heat capacity of the system at constant pressure. In practice, there are heat losses through convection, conduction and radiation that affect the temperature change observed.

There are three main types of thermal chemical sensors that have been miniaturised, the thermistor (§5.4.2), the pellistor (§5.7.2) and the thermopile (§5.2). Figure 9.17 shows the general arrangement of a thermistor device. The

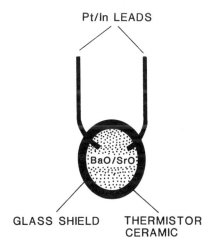

Pt/In LEADS

BaO/SrO

GLASS SHIELD   THERMISTOR CERAMIC

Figure 9.17 Schematic diagram of a chemically-sensitive thermistor.

centre is made of a high temperature sintered oxide (e.g. BaO/SrO) miniature bead with its resistance measured via two metallic (e.g. Pt/In) leads. Thermistors can have either a negative or positive temperature coefficient of resistance (TCR) and cover a typical temperature range of $-80°C$ to $+350°C$ with resistance from 100 Ω to 1 MΩ.

The relative resistance change, or thermistor coefficient, α_r of a thermistor is given by

$$\alpha_r = \frac{1}{R}\frac{dR}{dT} = \pm\frac{E_g}{kT^2} \qquad (9.11)$$

where E_g is the band-gap of the ceramic oxide semiconductor. The TCR is typically $\pm 5\%/°C$, and the glass coating makes the thermistor an extremely stable (*ca.* $\pm 0.05°C/yr$), low cost and chemically inert sensor probe. The small size of microbead thermistors provides a low thermal time constant (≤ 1 s).

Thermistors are commonly used to monitor enzyme-catalysed reactions as an alternative to an ISFET (§9.5.2) or MOS capacitor (§9.3). Table 9.5 shows the enthalpies of some common enzyme-catalysed reactions.

Table 9.5 Molar enthalpies of some common biosensing reactions.

Analyte	Enzymes	$-\Delta E_h$ (kJ/mol)
Cholesterol	Cholesterol oxidase	52.9
Glucose	Glucose oxidase (GOx)	80.0
H_2O_2	Catalase	100.4
Lactate	Lactate oxidase	-
Urea	Urease	6.6
Lipids	Lipase	-
Peptides	Trypsin	-
ATP	ATPase	-

A particularly important example is the glucose enzyme thermistor in which there are two reactions. First, the glucose is converted to gluconic acid by GOx and produces H_2O_2. Then the H_2O_2 reacts to produce water and oxygen:

$$\beta\text{-}D\text{-glucose} + H_2O + O_2 \xrightarrow{\text{GOx}} H_2O_2 + D\text{-gluconic acid}; \Delta E_{h1}$$

$$H_2O_2 \xrightarrow{\text{Catalase}} \tfrac{1}{2}O_2 + H_2O; \Delta E_{h2} \qquad (9.12)$$

The total change in enthalpy ($\Delta E_{h1} + \Delta E_{h2}$) can be measured using two thermistors in which the second one acts as a reference resistor in a Wheatstone bridge arrangement [§2.2]. Figure 9.18 shows the typical response of a glucose thermistor at different H_2O_2 concentrations.

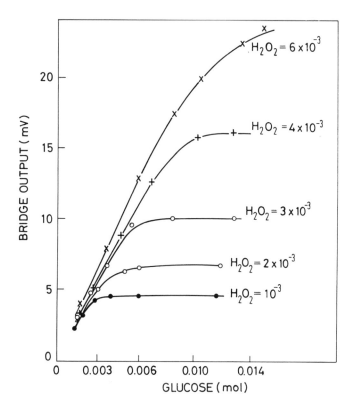

Figure 9.18   Typical response of a glucose thermistor at various molarities of H_2O_2 in an active bridge circuit.

9.6.2 Pellistor gas sensors

The pellistor is a miniature calorimeter (§5.7.2) used to measure the heat liberated on the oxidation of a combustible gas. The design was patented in the early 1970s by EEV (UK) and consists of a thin platinum coil surrounded by a porous refractory oxide bead (e.g. Al_2O_3) through which a catalytic metal is dispersed (e.g. platinum). The principle of the pellistor is similar to the metal oxide chemoresistors (§9.2.2), except that the temperature rather than conductance is measured [9.19]. The platinum coil both heats the catalytic bead to its operating temperature and acts as one element in a Wheatstone bridge (§2.2.1). The operating principle of the pellistor results in an ability to detect a wide range of combustible gases but with a low specificity. Table 9.6 lists some of the common gases that a pellistor may be used to detect.

Although pellistors are able to measure the lower (LEL) and higher (HEL) explosive levels of common combustible gases, they suffer from two problems. First, the porous oxide can be poisoned by silicones, and some chlorinated or sulphonated gases. Poison-resistant materials have been developed to overcome this problem. Secondly, the power consumption of the standard pellistor devices is rather high for portable instruments (~1 W). Recently, commercial beads have been commercially produced (~300 μm diameter) with much lower power consumptions.

Table 9.6 Some combustible gases which a pellistor detects.

Gas Property	Ethane	Benzene	n-Butane	CO	CH_4	H_2
Ignition temp. (°C)	305	560	365	605	595	560
LEL (%)	1.5 to 2.5	1.2 to 1.4	1.5 to 2.0	12.5	5.0 to 5.3	4.0
HEL (%)	81 to 100	8.0	8.5	74.2	14 to 15	75.6

A pellistor sensor has also been fabricated using silicon micromachining technology (Chapter 3). Figure 9.19 shows the cross-section of a silicon planar pellistor as well as a conventional bead pellistor. The substrate consists of a LPCVD Si_3N_4 (150 nm) membrane onto which a catalytic metal layer has been deposited [9.20]. The small structure results in a device with a very fast (< 1 ms) time constant (the heat capacity is only 10^{-8} J/K) and a low power consumption (12 to 20 mW at 200°C). Figure 9.20 shows the output of this Si planar pellistor at various concentrations of hydrogen and carbon monoxide.

Micromachined catalytic gas (pellistor-type) and semiconducting gas (chemoresistive-type) sensors are potentially superior to conventionally-designed commercially available gas sensors. However, there are some difficulties with these microdevices. For example, the fabrication of highly stable thin films, at

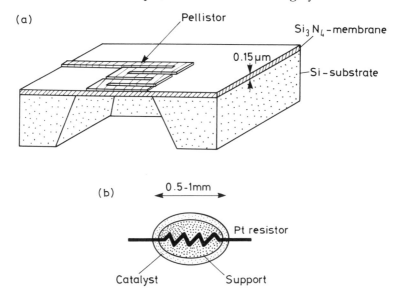

Figure 9.19 (a) A Si micropellistor [9.20], and (b) a conventional pellistor.

temperatures up to 700°C, is not proving to be easy even on a silicon nitride substrate. In addition, poisoning and interference by other gases can be a problem. New sensor designs and materials may overcome these problems in the next few years and generate a significant market.

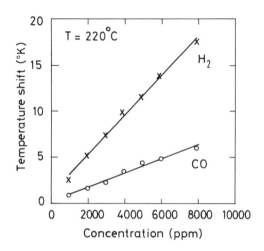

Figure 9.20 Response of a silicon micropellistor to H_2 and CO.

9.6.3 Thermocouple gas sensors

The heat liberated by a chemical or biochemical reaction can also be measured using a thermocouple or thermopile (§5.2). Micromachined microcalorimeters have been developed with a thermopile pick-up for the measurement in either a liquid or a gaseous environment. For example, a 0.4 µm thick LPCVD Si_3N_4 membrane has been used to form the substrate of an array of 56 n-type polysilicon-aluminium thermocouples to produce a gas-sensitive microcalorimeter [9.21]. The sensitivity of this device is about 10 mV/K and it is commercially available from Xensor Integration bv (Netherlands).

9.7 Mass-sensitive Chemical Microsensors

9.7.1 Piezoelectric devices

When a chemical species reacts with the sensing material, it often results in a change in the total mass. This small change in mass can be measured in a microbalance using either a piezoelectric or a surface acoustic wave device (§7.7); quartz is commonly used as a piezoelectric material.

Piezoelectric mass sensors (e.g. a quartz microbalance) have a high sensitivity when compared to other types of chemical sensors and are thus an attractive proposition. However, piezoelectric materials only function at a relatively low temperature (< 50°C) and so they are unsuitable for use with high-temperature oxide materials such as SnO_2.

The change in mass Δm is generally converted to a frequency shift Δf by an oscillator circuit (§7.7.2). The basic equation that relates them is known as the Sauerbrey equation,

$$\Delta f = -\frac{1}{\rho_m k_f} \cdot f_0^2 \cdot \frac{\Delta m}{A} \qquad (9.13)$$

where A is the crystal surface area, ρ_m is the density of the thin active coating on top of the piezoelectric substrate and k_f is a frequency constant. The physical pre-factor $(1/\rho_m k_f)$ is about 2.3×10^{-7} m^2 /Hz kg for a quartz plate.

The performance of piezoelectric mass (and SAW) chemical microsensors depends upon the functionality of the chemically-sensitive coating. A considerable variety of coatings have been developed, from spin-coated polymers to Langmuir-Blodgett lipid layers, to detect a wide range of chemicals. For example, Table 9.7 shows the effect of 19.3 ppm of strychnine (a bitter substance) and 19.3 ppm of β-ionone (an odour) on the adsorption mass (Δm) and partition coefficient P (ratio of the analyte concentration inside and outside the coating) with various active coatings on a Quartz Crystal Microbalance (QCM) [9.22].

Table 9.7 Adsorption masses and partition coefficients of 19.3 ppm strychnine or β-ionone for various active coatings on a QCM held at 45°C [9.22].

Coatings	Strychnine Δm (ng)	P	β-ionone Δm (ng)	P
Uncoated	2 ± 10	10	2 ± 10	10
2C$_{18}$N$^+$2C$_1$/PSS$^-$	533 ± 10	2,700	610 ± 10	3,050
DMPE^2	560 ± 10	2,800	540 ± 10	2,700
Poly(vinyl alcohol)	4 ± 10	18	4 ± 10	19
Poly(methyl glutamate)	5 ± 10	25	6 ± 10	30
Poly(styrene)	7 ± 10	35	7 ± 10	35
Bovine plasma albumin^3	5 ± 10	25	6 ± 10	30
Keratin	7 ± 10	35	6 ± 10	30

The selectivity of lipid-coated QCMs is mainly determined by the value of the partition coefficient which in turn relates to the mass, shape and dielectric constant

---

[2] Multibilayer cast film of dimylistoylphosphatidylethanolamine.
[3] Cross-linked with glutalaldehyde to immobilise on the QCM.

of the chemical molecule being sensed. Thus, the careful chemical design of the coatings ultimately determines the performance of this type of chemical microsensor.

9.7.2 SAW devices

SAW devices (§7.7.3) can be made at a relatively low cost using microlithographic and precision microengineering techniques (Chapters 3 and 4). Like a QCM, a SAW device can be coated with a thin soft polymer film and employed as a (bio)chemical microsensor. A SAW microsensor differs from a QCM in several ways. Firstly, the acoustic wave propagates near the crystal surface rather than through the bulk. Consequently, the Sauerbrey equation is replaced by the equation below [9.23],

$$\Delta f = (k_1 + k_2) f_0^2 \Delta(\rho_m d) \tag{9.14}$$

where $(\rho_m d)$ is the mass per unit area (d is the thickness of the coating and ρ_m is its density) which changes with measurand, and k_1 and k_2 are material constants (e.g. k_1 is -9.33×10^{-8} m^2 s/kg and k_2 is -4.16×10^{-8} m^2 s/kg for Y-cut quartz). Secondly, SAW microsensors generally work at much high frequencies ($\sim$ GHz) than a QCM ($\sim$10 MHz). This results in a higher mass resolution which is down to about 1 pg of material. Thirdly, SAW microsensors can be made much smaller than a QCM (at the same sensitivity) and are thus less expensive to manufacture.

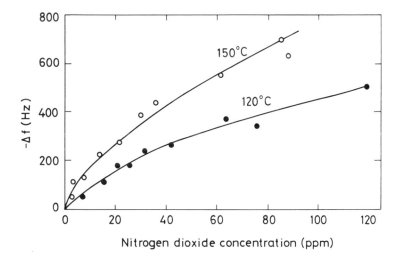

Figure 9.21 Response of a SAW chemical microsensor with a metal-free Pc coating [9.24].

Considerable work has been published on the measurement of inorganic gases (e.g. NO_2, H_2, H_2S, SO_2) and organic gases and vapours (e.g. CH_4, C_6H_6, C_2H_5OH). A good review article may be found in reference [9.24]. The choice of chemically-sensitive layer once again determines the selectivity of the sensor. Figure 9.21 shows the sensitivity of a SAW microsensor to NO_2 with metal-free Pc layers held at 120°C and 150°C [9.24]. The main problems with SAW microsensors are a poor long-term stability, high dependence on temperature and a high sensitivity to humidity. However, the development of intelligent compensation signal processing techniques (Chapter 11) may help ameliorate some of these shortcomings.

9.8 Applications of Chemical Microsensors

There is a considerable market for chemical sensors, although the market is smaller than that for other types of sensors at present. The Japanese market for gas/humidity solid-state sensors in 1987 (§1.5) was only valued at £7 million. However, this is an underestimate of the potential market in chemical sensors should microsensor technology become more firmly established. The most important fields of application are summarised in Table 9.8 together with some practical examples [9.1].

Table 9.8 Typical applications of chemical sensors.

Field of application	Typical example
Automotive	Engine control, air quality in car, emission
Aerospace	Engine control, air quality in cabin, emission
Agriculture	Fertiliser and pesticide control
Chemical analysis	Laboratory testing of materials
Safety (fire)	Fire warnings in mines, buildings, houses etc.
Process control	Production of chemicals, foodstuffs, etc.
Environmental monitoring	Detection of pollutants in air, water and soil
Medicine	Anaesthetic gases, diagnostics, biochemistry
Customs	Illegal and dangerous substances (explosives)
Quality control	Smell/flavour of drinks, foodstuffs, tobacco

Two application areas that are becoming increasingly important are the environmental monitoring/control of chemical substances, and medical diagnostics/practice. Both of these applications require quite sophisticated sensor systems and data processing methods. Consequently, there is a need to develop intelligent sensor array technology more rapidly to meet fully these requirements (§12.5).

Table 9.9 shows the world-wide market for various types[4] of chemical sensor in these two fields in 1988 and the predicted value for the year 2000. It may be seen from the table that the world-wide market in 1988 was £79 million of which £20 million was environment-related and £59 million medically-related.

Table 9.9 World-wide market for chemical sensors in 1988 and 2000.

Type	Field	1988 (£M)	2000 (£M)
Electrochemical	Environmental (gas/water)	17	97
	Medical (diagnostics and control)	19	70
Semiconducting	Environmental (gas/water)	3	90
	Medical	-	-
pH Sensors	Medical (diagnostics and control)	6	22
ISFETs	Medical (control)	3	13
Biosensors	Environmental (gas/water)	-	21
	Medical (diagnostics and control)	31	387

By the year 2000, the market is expected to grow by a factor of nearly ten to £700 million (£208 million environmental and £492 million medical). This represents an enormous growth in the market with the biosensor type showing the largest growth.

Suggested further reading

Janata J: *Principles of Chemical Sensors* (1989). Published by Plenum Press, New York. ISBN 0-306-43183-1. (317 pages. Good overview of chemical sensors)

Turner APR, Karube I and Wilson W (eds.): *Biosensors: Fundamentals and Applications* (1989). Published by Oxford Science Publications, Oxford, UK. ISBN 0-19-854745-5. (770 pages. Comprehensive review of biosensors)

References

9.1 Göpel W, Jones TA, Kleitz M, Lundström I and Seiyama T (eds.) (1991) *Sensors: A Comprehensive Survey, Vol. 2/3: Chemical and Biochemical Sensors*, VCH Publishers, Weinheim, Germany, 716 pp.

9.2 Heiland G and Kohl D (1985) Problems and possibilities of oxidic and organic semiconductor gas sensors. *Sensors and Actuators*, 8, 227-233.

9.3 *Technical Reference Manual*, Figaro Engineering Inc., Osaka, Japan, February 1992, 329 pp.

---

[4] Some devices are about to be miniaturised (e.g. electrochemical).

9.4 Corcoran P, Shurmer HV and Gardner JW (1993) Integrated tin oxide sensors of low power consumption for use in gas and odour detection. *Sensors and Actuators B*, **15-16**, 32-57.

9.5 Wright JD (1989) Gas adsorption on phthalocyanines and its effects on electrical properties. *Progress in Surface Science*, **31**, 1-60.

9.6 Cranny AWJ and Atkinson JK (1990) An investigation into the viability of screen printed organic semiconductor compounds as gas sensors. *Inst. Phys. Conf. Ser.* **No. 111**, 345-354.

9.7 Tredgold RH, Young MCJ, Hodge P and Hoofar A (1985) Gas sensors made from Langmuir-Blodgett films of porphyrins. *IEE Proc. I.*, **132**, 151-6.

9.8 Bott B and Jones TA (1984) A highly sensitive NO_2 sensor based on electrical conductivity changes in phthalocyanine films. *Sensors and Actuators*, **5**, 43-53.

9.9 Gardner JW and Bartlett PN (1991) Potential application of electropolymerised thin organic films in nanotechnology. *Nanotechnology*, **2**, 19-33.

9.10 Miasik JJ, Hooper A and Tofield BC (1986) Conducting polymer gas sensors. *J. Chem. Soc. Trans.*, **1(82)**, 1117-1126.

9.11 Bartlett PN and Ling-Chung SK (1989) Conducting polymer gas sensors. Part II. Response of poly(pyrrole) to methanol vapour. *Sensors and Actuators*, **19**, 141-150.

9.12 Lundström I and Svensson C (1985) Gas-sensitive metal gate semiconductor devices, in *Solid-state Chemical Sensors* (eds. J. Janata and R.J. Huber), Academic Press, New York, pp. 1-63.

9.13 Blackwood D and Josowicz M (1991) Work function and spectroscopic studies of interactions between conducting polymers and organic vapours. *J. Phys. Chem.*, **95**, 493-502.

9.14 Tan TT, Gardner JW, Farrington J and Bartlett PN (1993) Electronic properties of metal-poly(pyrrole) junctions. *Int. J. Electronics* (in press).

9.15 Assadi A, Spetz A, Willander M, Svensson C, Lundström I and Inganas O (1994) Interaction of planar polymer Schottky barrier diodes with gaseous substances. *Sensors and Actuators* (in press).

9.16 Lundström I, Shivaraman S, Svensson C and Lundkuist L (1975) A hydrogen sensitive MOS field-effect transistor. *Appl. Phys. Lett.*, **26**, 55-57.

9.17 Bergveld P (1970) Development of an ion-sensitive solid-state device for neurophysiological measurements. *IEEE Trans. Biomed. Eng.*, **19**, 70-71.

9.18 Bergveld P (1989) Exploiting the dynamic properties of FET-based chemical sensors. *J. Phys. E.: Sci. Instrum.*, **22**, 678-683.

9.19 Moseley PT and Tofield BS (eds.) (1987) *Solid State Gas Sensors*, IOP Publishing Ltd, Bristol, UK, 245 pp.

9.20 Gall M (1991) The Si planar pellistor: a low power pellistor sensor in Si thin-film technology. *Sensors and Actuators B*, **4**, 533-538.

9.21 van Herwaarden AW, Sarro PM, Gardner JW and Bataillard P (1993) Microcalorimeters for bio(chemical) measurements in gases and liquids. *Technical Digest of Transducers '93, Yokohama, Japan, June 1993*.

9.22 Okahata Y and Ebato H (1992) Detection of bioactive compounds using a lipid-coated quartz-crystal microbalance. *Trends in Analytical Chem.*, **11**, 344-354.

9.23 Wohltjen H (1984) Mechanisms of operation and design considerations for surface acoustic wave device vapour sensors. *Sensors and Actuators*, **5**, 307-325.

9.24 Fox CG and Alder JF (1992) Surface acoustic wave sensors for atmospheric gas monitoring, in *Techniques and Mechanisms in Gas Sensing* (eds. P.T. Moseley, J.O.W. Norris and D.E. Williams), Adam Hilger, Bristol, UK, pp. 325-345.

Problems

9.1 The number of charge carriers produced when sintered porous tin oxide reacts with a combustible gas cannot account for the large changes that occur in its bulk resistivity. Explain why this is the case. From what basic physical model do you think that equation (9.6) has been derived?

9.2 Are chemical microsensors in general linear or non-linear sensors? Explain the underlying principle or principles that led to your conclusion.

9.3 What frequency shift would you expect to see in a SAW microsensor in the presence of 1 ppb of ethanol? You may assume a centre frequency f_0 of 1.1 GHz for an active material of Pc (10 nm thick) and use typical values for any other parameters that you need.

9.4 Why do you think that chemical microsensors have not enjoyed the same commercial success of other classes of microsensors. How might these deficiencies be overcome? Are there any exceptions to this general rule? If so, name them.

10. Microsensor Performance

Objectives

☐ To identify key microsensor performance parameters

☐ To introduce the topic of microelectronic device reliability

☐ To consider the effectiveness of microsensor systems

10.1 Basic Considerations and Definitions

The suitability of a sensor for a particular application is essentially determined by its characteristic performance. Different applications may require fundamentally different sensor performances. For example, the temperature of a room does not normally need to be measured very precisely and so failure of a temperature, if it is not part of a control system, may be of little consequence. In this case, a low cost silicon thermometer, such as a thermotransistor (§5.6), is fit for the desired purpose. However, a microsensor situated inside a heart pacemaker to measure an erratic heart-beat must be extremely reliable, because its failure could lead to the death of its user. In this second example, the microsensor is said to be safety-critical.

The overall performance of a microsensor or microsensor system can be described by a parameter called the (system) *effectiveness*. There are essentially three factors that contribute to the effectiveness of a microsensor, see Figure 10.1.

First, the *capability* of a microsensor is its capacity to perform the desired function under predefined conditions. For example, consider the requirement to detect low levels of methane gas in air. In order to do this there must be a technical specification against which the microsensor's performance (i.e. its capability) can be measured and thus defined. Table 10.1 lists some of the important microsensor capability parameters that need to be defined for its use, e.g. as a pellistor-type gas microsensor (§9.6.2). Some of these parameters have already been discussed in an earlier chapter (§1.3) for the ideal microsensor.

250

Table 10.1 Some key sensor parameters that define its capability.

Operating specification	Pellistor example
Measurement principle	Calorimetric
Measurand	Concentration of methane
Measurement medium	Air
Measurement range	0 to 1,000 ppm
Sensitivity	5 mV/ppm
Resolution	± 10 ppm
Accuracy	± 1%
Response time	< 1 s
Recovery time	< 2 s
Linearity of output signal	1%
Maximum sampling rate	0.1 Hz
Temperature range	−20 to + 60°C
Power consumption	700 mW
Other	Dimensions, weight etc.

The second factor determining the effectiveness of a microsensor is its inherent *reliability*. Reliability is defined by BS 4778 as the ability of an item to perform a required function under stated conditions for a stated period of time [10.1]. Thus, the reliability of a microsensor is its ability to stay within its technical specification over a prescribed period of time. In this respect, the reliability of a sensor is a measure of its output holding to a calibration and so producing acceptable output signals. Table 10.2 lists some of the key parameters that determine the reliability of a microsensor. A pellistor is used once again to illustrate the details.

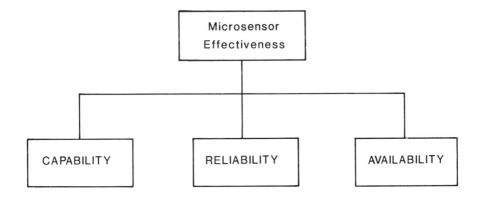

Figure 10.1 Effectiveness of a microsensor or microsensor system.

Table 10.2 Some key parameters that determine the reliability of a microsensor.

Operating specification	Pellistor example
Method of calibration	Digital LUT stored in electronic memory
Class/type of sensor	Chemical calorimetric
Type of interface circuitry	Wheatstone bridge and voltage amplifier
Calibration drift	Drift in IC parameters (%/h)
Calibration interval	(h)
Drift of zero point	Variation in offset (%/h)
Drift of sensitivity	(mV/ppm)/h
Life-time	(h)
Contamination	Poisoned by silicones
Interference	Degraded signal due to humidity variation

A more detailed discussion of the reliability of microelectronic devices is given later (§10.3).

Finally, the *availability* of a microsensor is also of importance in many practical applications. The availability is defined as the probability that an item is capable of performing a required function when required to do so [10.1]. The availability of a microsensor is in fact defined by its failure-rate and its repair-rate (e.g. its maintainability). Table 10.3 lists some of the key parameters that directly determine the maintainability of a microsensor and hence its availability. In the examples discussed above, the availability of a room thermometer may be low, but that of a heart pacemaker or explosive gas detector should be very high (e.g. close to 100%).

Table 10.3 Some key microsensor parameters that define its maintainability.

Specification	Pellistor example
Class/type of sensor	Chemical calorimetric
Type of interface	Wheatstone bridge
Calibration intervals	1 yr
Equipment needed for calibration	Laboratory test facility
Maintenance intervals	3 months
Ease of diagnostics/repair	Use of error signals, test-points etc.
Battery life	48 h
Delivery of spare-parts	1 day

The necessary level of microsensor effectiveness is essentially determined by its application and hence by the user. Consequently, a microsensor will be designed by an engineer so that its effectiveness matches the requirements of the customer, i.e. functionality, cost, quality, size, weight, speed etc. In some cases there will also be safety-critical customer (or user) specifications to be met.

We now consider in more detail some of the procedures and parameters which determine the effectiveness (i.e. performance) of a microsensor or microsensor system.

10.2 Calibration of Microsensors

10.2.1 Definitions

There is a natural variability of the fundamental parameters and properties of microsensor materials (e.g. Young's modulus) that relates to the manufacturing processes used (e.g. silicon micromachining). This in turn leads to some variation in the performance of the microsensor. If this variation is large compared to the tolerance permitted then either some calibration technique must be employed or the design must be re-evaluated at a more intelligent level (§11.1).

Calibration is the process that is carried out to ensure that a microsensor conforms to a known standard within a specified tolerance. For example, the micromachining of silicon microsensors often requires the chemical etching of both active and passive materials (Chapter 4). As it can be difficult to control the etching process, small variations in the dimensions of a microdevice normally occur. These can then lead to a variation in perhaps the device capacitance or resistance, and hence the response of a microsensor.

Improvements in process control can often lead to a reduction in the parametric variability but this may be an expensive option when very low tolerances are required. Thus, a calibration technique is often the preferred way of assuring the performance of a microsensor.

A calibration is a test during which known values of the measurand are applied to a sensor and the output readings are recorded (i.e. set-points). The resulting calibration record may be plotted to produce a calibration curve. The calibration curve (or table) enables the user to determine the error characteristics of the microsensor.

Some common microsensor characteristics which are determined from a calibration curve are:

- response;
- operating range;
- linearity (if appropriate);
- hysteresis;
- repeatability;
- zero off-set;
- sensitivity.

The static calibration errors are usually quoted in units of the operating range (e.g. % FSO) whereas a dynamic calibration error would be in units of frequency (e.g. Hz).

10.2.2 Error correction

The calibration data are often used to reduce the output error characteristic of a microsensor. A simple modification may be made, for example, using a trimming potentiometer within an analogue electronic circuit (or the laser trimming of a resistor in an IC). In addition, a zero off-set can usually be corrected by adjusting the null off-set of an op-amp in a read-out circuit. Further, the gain of a circuit can be set by a feed-back trimming resistor. This procedure is useful when there are only a few reference points to calibrate. For instance, the output from a linear sensor (§1.3) can be calibrated with only two set-points which provide the slope and intercept. In cases where a set of calibration points is required to define a non-linear microsensor characteristic, it may be easier to use digital memory (e.g. store the calibration points in a LUT in non-volatile memory) or computationally using a microprocessor to calculate a scientific calibration function (e.g. a high order polynomial). An advantage of using the digital storage method is that it is inexpensive but, in some applications, it may be necessary to use a mathematical model to obtain a very precise calibration (e.g. a precision pressure microsensor).

Figure 10.2 shows the observed voltage output V_{out} of a silicon thermal microbridge flow-rate microsensor (§7.4.2) as a function of volumetric flow-rates used for calibration. These calibration points can be used to determine an analytical approximation (frequently a polynomial) to the observed microsensor output. An off-set voltage of 0.980 V has been recorded. A linear fit (Figure 10.2(a)) to the experimental data is seen to be rather poor with a sample linear regression error ε of 18.4%. Table 10.4 shows the improvement in the mathematical fit to the data as the degree n of the polynomial increases from one (linear) to seven. The quadratic (n=2) and quartic (n=4) fits are shown in Figure 10.2(b) and (c), respectively.

Table 10.4 Polynomial fitting of calibration data.

Degree n	Predicted zero	Zero error (%)	Correlation $\wp$	Regression error ε (%)
1	1.548	58.0	0.9779	18.4
2	1.228	25.3	0.9962	7.84
3	1.106	12.9	0.9986	4.92
4	1.020	4.1	0.9998	1.86
5	0.994	1.4	0.9999	1.02
6	0.992	1.2	0.9999	1.04
7	0.987	0.7	0.9999	0.94

Clearly, the polynomial fit to the data converges onto the observed data at higher orders. Although there is a good fit to the data at n=5, above it there appears to be an asymptotic value of about 1.0. This may be associated with the error in the calibration points themselves.

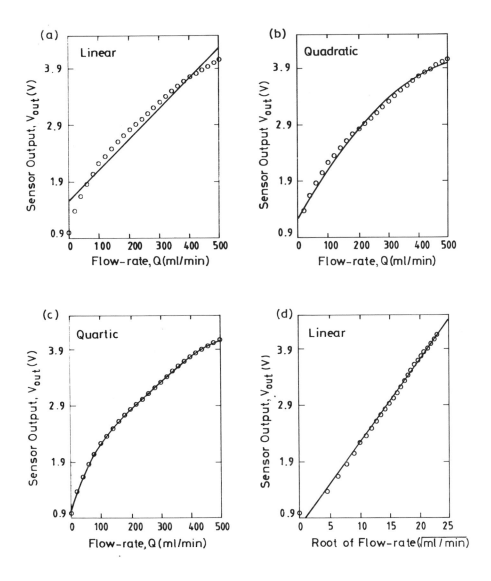

Figure 10.2 Various calibration plots for the output of a thermal microbridge flow-rate microsensor using a polynomial curve fitting technique.

Knowing the properties of the microsensor technology often permits a more direct and efficient calibration process. Thus, for a thermal flow-rate microsensor, the output voltage is approximately proportional to the square-root of the flow-rate. Thus, plotting the data accordingly produces a good linear fit to the data over the region of 5 to 500 ml/min. Figure 10.2(d) shows the linear fit with a sample correlation coefficient $\wp$ of 0.9980 and error ε of 5.5% (including the poor zero point). The zero point looks rather high and so calibration of the microsensor between 0 and 5 ml/min may be impractical.

More sophisticated calibration methods can be used to compensate for temperature or humidity and some discussion is made on this topic in the chapter on smart sensors (§11.4).

10.3 Reliability of Microelectronic Devices

10.3.1 Introduction

The reliability $\mathfrak{R}_i(t)$ of a device i is, as stated above, the probability that the device will perform satisfactorily over a prescribed period of time t. A common cause of failure is when the load applied to the device (e.g. gate voltage, stress) exceeds its material or component strength (e.g. dielectric breakdown strength, Young's modulus). For example, when the load voltage across an IC exceeds its specified operating value then a stress failure will occur. As there is stochastic variability in both the applied load L and the material strength S, the load and strength may be represented by probability distributions. Figure 10.3 shows a Gaussian representation of the load probability distribution $f_L(z)$ and strength probability distribution $f_S(z)$ as a function of stress z. The shaded area of overlap indicates the probability of the device failing as soon as the load is applied. It is clearly important to operate the device so that the average values of load $\bar{z}_L$ and strength $\bar{z}_S$ are well separated. Three key parameters are often used to define the characteristic load-strength behaviour of a device,

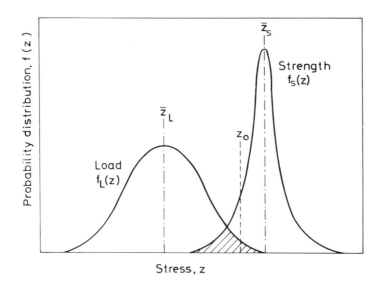

Figure 10.3 Gaussian probability distributions of load and strength for a device.

$$\text{Safety-Factor (SF)} = \frac{\bar{z}_S}{\bar{z}_L} \tag{10.1}$$

$$\text{Safety-Margin (SM)} = \frac{(\bar{z}_S - \bar{z}_L)}{\sqrt{\xi_S^2 + \xi_L^2}} \tag{10.2}$$

$$\text{Loading Roughness (LR)} = \frac{\xi_L}{\sqrt{\xi_S^2 + \xi_L^2}} \tag{10.3}$$

Clearly, the standard deviations of the load ξ_L and strength ξ_S affect the failure probability and this is described by the safety-margin. The loading roughness defines the variation in load compared to the variation in strength and is ideally kept below $1/\sqrt{2}$.

The reliability $\Re$ of a device can now be calculated by,

$$\Re = \int_{-\infty}^{+\infty} f_S(z_S) \int_{-\infty}^{z_o} f_L(z_L) dz_L dz_S ; \ z_S > z_L \tag{10.4}$$

where z_o is a dummy variable in the integration. Equation (10.4) can be readily solved for Gaussian probability distributions in load and strength, and produces the simple result,

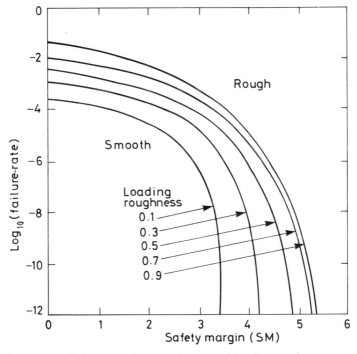

Figure 10.4 Failure-rate (per application of load) vs. safety margin of a device with normally-distributed load and strength distributions [10.2]. (Redrawn by permission of Macmillan Press Ltd, UK.)

$$\Re = f_\Phi \left(\frac{(\bar{z}_S - \bar{z}_L)}{\sqrt{\xi_S^2 + \xi_L^2}} \right) = f_\Phi \left(\frac{(\bar{z}_S - \bar{z}_L)}{\xi_L} LR \right) = f_\Phi (SM) \qquad (10.5)$$

where f_Φ is the cumulative probability of the Gaussian (i.e. normal) distribution.

Using equation (10.5), the failure-rate (probability of failure occurring when a load is applied) can be plotted against the safety-margin at various values of the loading roughness as shown in Figure 10.4 [10.2]. At a high safety-margin and low loading roughness, the probability of failure is low. For example, a device can withstand one million applications of the load when the safety-margin is 3 and the loading roughness is 0.1.

The characteristic regions of the failure-rate diagram at fixed loading roughness may be categorised as shown in Figure 10.5 [10.2]. In region 1 the failure-rate is unacceptably high, in region 2 the failure-rate is very sensitive to a change in the parameters, but in region 3 the design is virtually defect-free and hence intrinsically safe.

Ideally, the reliability of a device is close to unity after the weak devices have failed. In practice, however, the load-strength distributions are time-dependent. Examples of physical mechanisms that produce a time-dependent failure-rate are ion migration, crack formation and growth, creep, corrosion and, in mechanical devices, wear. The time-dependent failure-rate λ_f of a device is related to its reliability by,

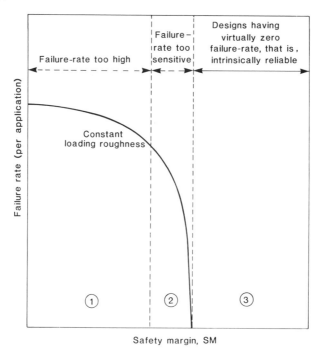

Figure 10.5 Characteristic regions of the failure-rate (per load) diagram [10.2]. (Redrawn by permission of Macmillan Press Ltd, UK.)

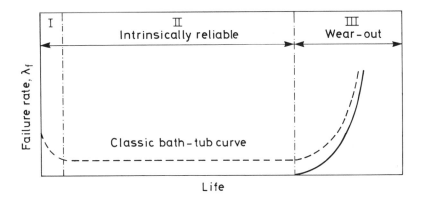

Figure 10.6 Characteristic regions of the failure-rate bath-tub distribution [10.2]. (Redrawn by permission of Macmillan Press Ltd, UK.)

$$\lambda_{\mathrm{f}} = \frac{-1}{\Re}\left(\frac{\mathrm{d}\Re}{\mathrm{d}t}\right) \qquad (10.6)$$

The characteristic life of some devices (e.g. mechanical) can be often described by the so-called "bath-tub" distribution, see Figure 10.6. The failure-rate diagram has once again three characteristic regions but now the failure-rate is plotted against the life-time of the device (dashed line). In the first region (I), the weak items quickly fail in a process known as infant mortality. A 100% inspection (i.e. pre-testing) of the devices or a deliberate burn-in (i.e. overstressing) process can remove this failure mode (solid line). In the second region (II), the failure mechanism is random in nature and so the failure-rate is constant and independent of time. This region determines the practical life-time of the device during which it is intrinsically reliable. Finally, the failure-rate increases in region III as the devices start to wear out due to ageing. Failure mechanisms of this sort are more pronounced in mechanical devices where fatigue, corrosion, surface degradation, defastening, etc. leads to failure. Thus the failure-rate of a device may decrease, be constant, or increase with time depending upon the type and age of the device. All three probability distributions can be described by the Weibull distribution $f_{\mathrm{w}}(t)$,

$$f_{\mathrm{w}}(t) = \left(\frac{\beta_{\mathrm{w}}}{\eta_{\mathrm{w}}}\right)\left[\frac{(t-t_0)}{\eta_{\mathrm{w}}}\right]^{\beta_{\mathrm{w}}-1} \qquad (10.7)$$

where t_0 is the time at which failures start, η_{w} is a measure of the characteristic life of the device, and β_{w} determines the characteristic shape of the curve. When β_{w} is less than 1 the failure-rate falls with age (i.e. infant mortality), when β_{w} equals 1 a

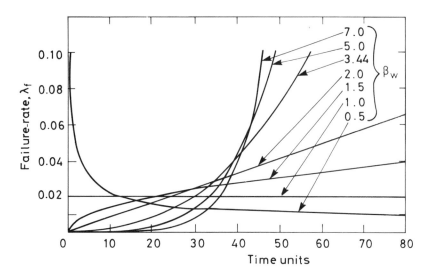

Figure 10.7 Characteristic failure-rate curves generated by the Weibull probability distribution [10.2]. (Redrawn by permission of Macmillan Press Ltd, UK.)

constant failure-rate occurs (i.e. a random failure mechanism) and when β_w is greater than 1 the failure-rate increases with age (i.e. a wear mechanism).

Figure 10.7 shows the failure-rate (sometimes called the hazard function) corresponding to values of the Weibull parameter β_w which cover all three regions of the bath-tub distribution. The normal distribution is equivalent to a shape parameter value of 3.44. Special graph paper is commercially available to plot the device life-times from which the values of the characteristic life η_w (mean time to fail) and the shape parameter β_w can be read off.

10.3.2 Electronic device reliability

The mechanisms of failure that occur in electronic devices are generally different from those that occur in mechanical devices. Considerably more is known about the failure of microelectronic devices than micromechanical devices. As microsensors may employ an electronic sensing element (e.g. a phototransistor) and an electronic read-out circuit, electronic reliability is of fundamental importance to the field of microsensors and microsensor systems.

Large electronic devices (e.g. a solid-state switch) generally have a lower failure-rate than their mechanical counterparts (e.g. a reed relay) and it is common for the failure-rate of such electronic devices to decrease with time. However, miniaturisation and production of very small complex microdevices may offset this. For example, aluminium is commonly used as the metallisation layer in VLSI devices. The use of a passivation layer can lead to a higher failure-rate if the coating generates a stress (tensile or compressive). Figure 10.8 shows a Weibull plot of the cumulative open-line failures $(1 - \Re(t))$ as a function of the ageing time of a 0.9 μm aluminium track on silicon at 295°C [10.3]. The higher failures of the

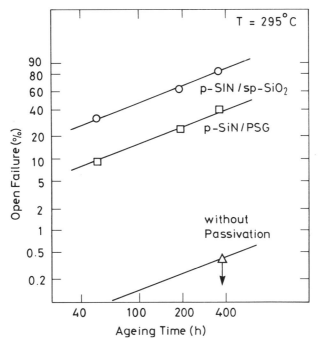

Figure 10.8   Weibull plot of cumulative open-circuit failures in aluminium tracks as a function of ageing time [10.3] (© 1993 IEEE).

stressed coatings show the importance of good practice in the processing of silicon devices. In this case polyimide passivation layers would be superior.

In microelectronic devices, the failures that are most likely to occur include the breakdown of a wire-bond to a junction, defective encapsulation, a defective semiconductor material and thermal runaway. Despite the example given above, microelectronic devices too usually have a decreasing failure-rate with time (i.e. β_w < 1) and this can be related to some fundamental parameters and properties of the device. Failure-rate models have been developed for application to discrete analogue electronic components (e.g. resistors, capacitors, transistors) and microelectronic devices [10.4, 10.5]. For semiconductor devices (e.g. an *n-p-n* transistor), the failure-rate λ_{es} is modelled by the following standard formula [10.4]:

$$\lambda_{es} = \lambda_b \left(\Pi_V \Pi_P \Pi_Q \Pi_C \Pi_E \Pi_A \right) \quad (/10^6 \, \text{h}) \tag{10.8}$$

where λ_b is the base failure-rate and is exponentially related to the operating temperature of the device, Π_V is a voltage-derating stress factor, Π_P is a transistor power rating factor, Π_Q is a component quality level that relates to the amount of screening carried out, Π_C is a complexity factor that depends on the number of transistors per package, Π_E is an environmental factor to take account of harsh operating conditions, and Π_A relates to the type of application (e.g. switching).

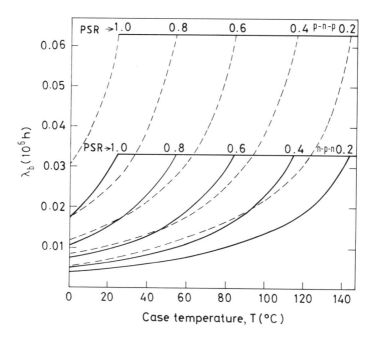

Figure 10.9 Base failure-rate of *n-p-n* and *p-n-p* semiconductor devices at various power stress ratios [10.3] (© 1993 IEEE).

Figure 10.9 shows the base failure-rate λ_b of both *n-p-n* and *p-n-p* transistors at various Power Stress Ratios (PSRs). In general, *n-p-n* transistors have a higher reliability than *p-n-p* transistors and where possible these should be used. Moreover, running transistors at lower loads (e.g. a PSR of 0.5) is clearly beneficial in terms of significantly lower failure-rates whether using *n-p-n* or *p-n-p* transistors.

The failure-rate model has been extended to cover most types of microelectronic devices [10.4]. The formula now becomes,

$$\lambda_{em} = \left(\Pi_T\Pi_V\Pi_{C1} + \Pi_E\left(\Pi_{C2} + \Pi_{C3}\right)\right)\Pi_Q\Pi_L \qquad (/10^6\,h) \qquad (10.9)$$

where Π_T is a temperature-dependent factor like λ_b, Π_V and Π_Q are defined as before, Π_{C1-C3} are complexity factors depending upon the bit count in memory devices and transistor count in linear devices, and Π_L is a learning factor which is one for mature technologies and up to 10 for relatively new technologies.

Figure 10.10 shows the effect of the junction temperature upon the base failure-rate Π_T of microelectronic devices [10.4]. Again the failure-rate rapidly increases with an increasing junction temperature. In addition it is apparent that TTL microelectronic technology has considerably lower failure-rates than MOS and linear technologies, and that plastic encapsulated devices are less reliable than hermetically sealed (metal case) microelectronic devices. The latter is due to the fact that hermetically-sealed devices are less sensitive to the environment and have better transient thermal properties.

The relative reliability of TTL and MOS technologies is also affected by the complexity factors Π_{C1} and Π_{C2}. Figure 10.11 shows the effect of the gate count upon both of these complexity parameters (Π_{C1} and Π_{C2}) for TTL and MOS. Although the precise failure-rate is a complex function of various load, materials and processing parameters, it is possible to make some general statements about the relative performance of these technology bases. Table 10.5 compares the relative merits of these technologies where it is evident that the mature TTL technology is often preferable [10.6].

Table 10.5 Reliability and applications of microcircuit devices [10.6]. (Reprinted by permission of John Wiley & Sons Ltd, © 1991.)

Technology base	Applications	Reliability aspects
Digital Silicon transistor-transistor logic (TTL)	The most common technology for digital (logic) functions; the basis of 5400 (military) and 7400 (commercial) device range. Available in standard, low power (L), high speed (H), and Schottky (S) (very high speed) versions.	Very reliable, mature technology.
Metal oxide silicon (MOS)	Used when very low power consumption is required, and speed is not critical, e.g. calculators, watches.	Susceptible to damage from electrostatic discharges. Low power dissipation leads to lower temperature. However, lower activation energy makes reliability more sensitive to temperature.
Linear analogue Silicon transistors and associated circuitry	Amplifiers, regulators, analogue-digital and digital-analogue converters, etc.	Usually very reliable. However, many types are not as standardised as for digital devices and there can therefore be more variability.
Failure effects for digital devices:	Stuck at (digital) 1 (SA1). Stuck at (digital) 0 (SA0). Unstable/intermittent. Out of tolerance parameter (speed, switching voltage level).	
Failure effects for linear devices:	Hardover positive or negative. Unstable/intermittent. No output. Parameter out of tolerance, e.g. gain.	

The level of device screening determines the quality factor Π_Q in equation (10.9). There are several screening levels defined in standards from the highest class A (military specification) to the lowest class D (commercial). Table 10.6 shows the screening that is carried out for the top three classes of A, B and C from the US and UK standards MIL-STD-883 and BS 9400 [10.5]. Each screen is effective against certain defects as tabulated. The amount of screening determines both the quality factor and the cost of the device.

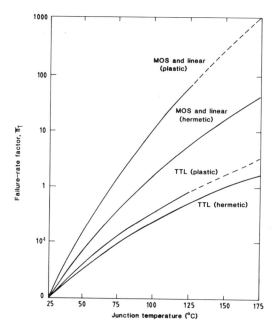

Figure 10.10 Base failure-rate of MOS/linear and TTL microelectronics devices as a function of junction temperature [10.6]. (Reprinted by permission of John Wiley & Sons Ltd, © 1991.)

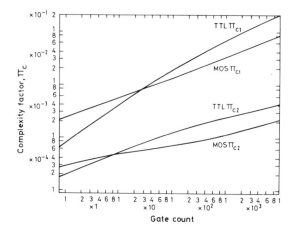

Figure 10.11 Effect of gate count upon the complexity factors in MOS/linear & TTL microelectronic device [10.6]. (Reprinted by permission of John Wiley & Sons Ltd, © 1991.)

Table 10.6 Microelectronic device screening requirements [10.6]. (Reprinted by permission of John Wiley & Sons Ltd, © 1991.)

Screen	Defects effective against	Screen level applicability (%) MIL-STD-883/BS 9400		
		A	B	C
Pre-encapsulation visual inspection (×30-200 magnification)	Bonds, contamination, chip surface defects, wire-bond positioning.	100	100	100
Stabilisation bake	Bulk silicon defects, metallisation defects, stabilises electrical parameters.	100	100	100
Temperature cycling	Package seal defects, weak bonds, cracked substrate.	100	100	100
Constant acceleration (20,000 *g*)	Chip adhesion, weak bonds, cracked substrate.	100	100	100
Leak tests	Package seal.	100	100	100
Electrical parameter tests (pre-burn-in)	Surface and metallisation defects. Bond failure. Contamination/particles.	100	100	100
Burn-in test (125°C with applied AC voltage stress)	Surface and metallisation defects. Weak bonds.	100 240 h	100 160 h	-
Electrical test (post burn-in)	Parameter drift.	100	100	-
X-ray	Particles, wire-bond position.	100	-	-

Table 10.7 shows the variation of the quality factor Π_Q with 8 different screening levels [10.5]. A higher class of screening produces a lower value of Π_Q and hence a lower failure-rate of the device. But this also leads to a substantial cost penalty of perhaps up to a factor of 20 for a class A device compared to a

Table 10.7 Microcircuit device screening effectiveness and cost [10.6]. (Reprinted by permission of John Wiley & Sons Ltd, © 1991.)

Screen level (MIL-HBK-217)	Specification	Π_Q	Cost factor
A	MIL-M-38510, Class S	0.5	8-20
B	As above, Class B	1.0	
B-1	MIL-STD-883 method 5004, Class B	3.0	4-6
B-2	Vendor equivalent of B-1	6.5	
C	MIL-M-38510, Class C	8.0	2-4
D	Commercial (i.e. no special screening), hermetic package	17.5	1
D-1	Commercial plastic encapsulated	35.0	0.5

standard class D commercial device. Consequently, the balance between cost and reliability is set according to the particular application that is required.

10.4 Microsensor Reliability

The overall reliability of a microsensor system $\mathfrak{R}_{sensor}$ is ideally the product of the reliabilities of the sensing element, the processor (e.g. interface electronics), and the actuator (see Figure 1.1) that is,

$$\mathfrak{R}_{sensor} = \mathfrak{R}_{element} \times \mathfrak{R}_{processor} \times \mathfrak{R}_{read-out} \qquad (10.10)$$

where the overall reliability is often dominated by the lowest reliability contribution which may, depending on the device, be the sensing element (e.g. a pellistor), the processor (e.g. a Wheatstone bridge circuit) or the read-out instrumentation (e.g. liquid crystal display). The reliability of the processor and read-out electronics can be determined from the standard models presented above.

In cases where the sensing element and processor are integrated to create a smart sensor (Chapter 11), then the two reliability terms are no longer

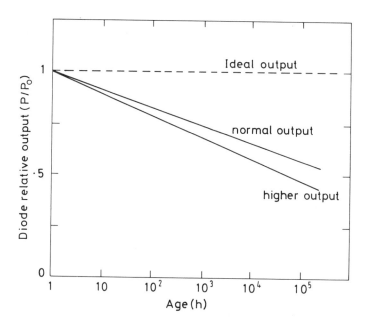

Figure 10.12 Degradation of the output of an IR photodiode over time.

independent and so a single term may be more appropriate. This is because the sensing element and processor now share the same processing technology and so can have a common mode of failure.

The reliability of a microsensor will also depend upon the complexity of the system. Consequently, array sensors (Chapter 12) which consist of many identical microsensors will have a lower reliability than individual discrete devices.

In practice, it is rather difficult to construct an exact failure-rate model for a microsensor because of the special processing methods (Chapter 4) used to deposit active sensing materials onto microelectronic substrates. However, appropriate models should be possible in the future for mass-produced microsensors.

It is quite common for the performance of a microsensor to degrade with time rather than to completely fail. This is normally due to a device parameter drifting or ageing so that the output is affected. For example, Figure 10.12 shows the output from an IR photodiode employed in an optical displacement microsensor (§8.3) as a function of time and there is a significant decrease of relative power output with time. This is due to the formation of microcracks and is a function of the device current. It is therefore often necessary to construct sensors that ameliorate the effect of parametric drift. Alternatively, parameter drift in sensors can be corrected either through the use of frequent calibration (i.e. regular maintenance) or by electronic compensation through intelligent microsensor design (Chapter 11).

Suggested further reading

Readers are referred to the following texts that provide some background information for this chapter on the reliability of microsensors:

O'Connor, PDT: *Practical Reliability Engineering* (1991). Published by Wiley & Sons Ltd, UK. ISBN 0-471-92902-6. (409 pages. Good introduction to the topic of reliability)

Norton, HN: *Handbook of Transducers* (1989). Published by Prentice-Hall, New Jersey, USA. ISBN 0-13-382599-X. (554 pages. Chapter 4 on Transducer Performance Tests is good background reading)

References

10.1   *Quality Vocabulary*, BS 4778 (1979), British Standards Institution, London.

10.2   Carter ADS (1986) *Mechanical Reliability*, Macmillan, London, UK, 492 pp.

10.3   Takeda E, Ikuzaki K, Katto H, Ohji Y, Hinode K, Hamada A, Sakuta T, Funabiki T and Sasaki T (1993) VLSI reliability challenges: from device physics to wafer scale systems. *Proc. IEEE*, **81**(5), 653-674.

10.4   *Reliability Prediction of Electronic Equipment*, MIL-HDBK-217 (1980), US standard.

10.5   Test methods and procedures for microelectronic devices, MIL-STD-883 or BS 9400.

10.6   O'Connor PTD (1991) *Practical Reliability Engineering*, John Wiley & Sons Ltd, UK, 409 pp.

Problems

10.1 What is a load-strength diagram, and why is it important in assessing the reliability of microsensors? By considering the loading roughness and safety-margin, sketch the load-strength diagrams to show (a) the ideal case for a reliable microsensor, (b) the case where the loading roughness is too high, and (c) the case where the device strength is too high. Which factor do you think is more undesirable in the application of a microsensor: variation in the load applied or variation in the device strength? Explain why.

10.2 An average load voltage of +10 V with a standard deviation of 1 V is applied to a silicon microsensor. The device is rated at an average strength of +12 V. The loading roughness has been calculated as 0.707 for a particular application. Calculate the safety-margin of the device used in this application.

10.3 A manufacturer uses IR light-emitting diodes in an optical displacement microsensor. Tests are made on a random sample of 15 devices and these record the age at which they fail (in hours) as 710; 1,540; 670; 970; 475; 780; 1,230; 680; 810; 370; 1,060; 890; 1,410 and 850. Consider the possible mechanism(s) of device failure, and thus explain why a Weibull model is appropriate. *(Do not perform a Weibull analysis of the data)*

10.4 Calculate the Weibull parameters for the data given in problem 10.3. What factors do you think will affect the validity of the results of your Weibull analysis?

11. Smart Sensors

Objectives

☐ To introduce the topic of integrated, smart and intelligent sensors

☐ To describe the functionality of intelligent sensors

☐ To give some examples of smart sensors

11.1 Integrated, Smart and Intelligent Sensors

The successful application of many types of solid-state sensor has stimulated the market but led to a demand for either a lower unit cost or an enhanced functionality to improve the market value. Both of these may be achieved through a higher level of device integration. In addition, more difficult sensing problems are now being studied which require a higher level of processing power (i.e. more intelligence) than is achievable from current sensors. Thus, the major thrust of today's research effort is to make sensors smaller and smarter than they are now.

The basic arrangement of a measurement system and its level of integration was discussed briefly in Chapter 1 (§1.1 and §1.6). In order to define more carefully the different types of intelligent sensor, it is necessary to subdivide the *processor* into a signal preprocessor (converter) and the main processing unit (e.g. a microprocessor). The signal preprocessor or converter carries out low level tasks such as amplification, filtering, or analogue-to-digital conversion (Chapter 2). Figure 11.1 illustrates the three levels of integration which make up sensor systems.

There are differing views at present over the precise definition of a smart sensor. One [11.1] proposes that a device onto which at least one sensing element and signal processing circuit has been integrated is a smart sensor. The problem with this definition is that most sensors would be called smart - even when they possess such a low level of intelligence that to call them intelligent is rather ambitious. Instead we will use the term *integrated sensor* to describe this type of

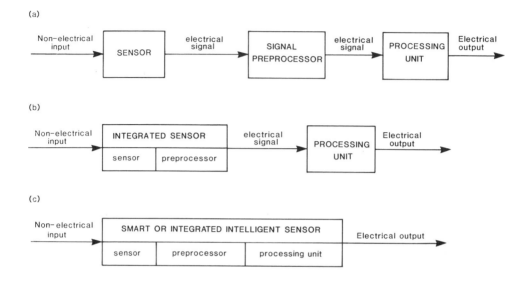

Figure 11.1 (a) A sensor system, (b) an integrated sensor in a sensor system, and (c) a smart sensor or integrated intelligent sensor.

low-level smart sensor where most of its preprocessor is integrated, see Figure 11.1(b). This revised definition is now consistent with that proposed more recently by Hauptmann [11.2]. The term *smart* is reserved to denote the integration, in part or full, of the main processing unit, see Figure 11.1(c), which adds intelligence.

There is little confusion in this practical definition because all smart sensors must be integrated and intelligent, while any sensor that has significant intelligence which has not been integrated can be called an *intelligent sensor*. The only definition that remains is that of an intelligent sensor. The definition proposed by Breckenbridge and Husson [11.3] takes some account of work in artificial intelligence and runs as follows:

> *"The sensor itself has a data processing function and automatic calibration/automatic compensation function, in which the sensor detects and eliminates abnormal values or exceptional values. It incorporates an algorithm, which is capable of being altered, and has a certain degree of memory function. Further desirable characteristics are that the sensor is coupled to other sensors, adapts to changes in environmental conditions, and has a discrimination function".*

This definition is rather long but does incorporate the essential functions required to define intelligence in a sensor. For the purposes of this book, an intelligent sensor must possess one or more of the following three features:

- perform a logical function;

- communicate with one or more other devices;
- make a decision using crisp or fuzzy data.

This definition is not only consistent with that proposed by Giachino [11.4] for a smart sensor but also distinguishes between an integrated sensor and a hybrid intelligent sensor.

In order for an intelligent sensor to perform a logical function it clearly requires some type of processing unit. The processing unit is most likely to be a microprocessor (e.g. Motorola 80HC11) but could be some type of programmable logic device. The integration of the processing unit with the sensor to make a smart sensor then requires a significant electronic design effort, perhaps through the use of Application Specific Integrated Circuit (ASIC) technology.

The ability of a sensor to communicate (i.e. exchange information with another device) is particularly useful. An intelligent sensor may be able to communicate with its operator and so provide valuable information about problems etc. Alternatively, an intelligent sensor may communicate with another device and so modify its own behaviour. This type of intelligence can, in its simplest form, provide a warning of abnormal operating conditions, or more cleverly provide a feedback control mechanism. Intelligent sensors may provide a good level of control such as via a PID digital controller. An intelligent sensor may have some form of high level adaptive control strategy which permits the control parameters to be automatically updated with time. The implementation of a sensor which can warn its user, or adapt to environmental conditions, requires some decision-making capability. Traditionally, sensors use parametric data to make a decision. For example, the signal from a thermodiode (§5.5) can be used to provide an overload protection to a device which exceeds its normal operating temperature. However, the more intelligent sensors of the future may use non-parametric methods such as artificial neural networks [11.5] or a so-called expert system which relies upon fuzzy data [11.6].

The definition of an intelligent sensor used here includes any sensor system

Table 11.1 Examples of intelligent sensor systems.

Sensor class	Principle of intelligent sensor	Application
Radiant	Monitors the spatial Fourier transform of the retroreflected light from a surface.	Machine tool control
Mechanical	Monitors the generation of acoustic noise by cracks in a metal.	Non-destructive testing
Mechanical	Effector sensors for handling objects.	Robotics
Chemical	Monitors air-intake and fuel combustion.	Engine control
Chemical	Monitors air condition for comfort and safety.	Buildings
Magnetic	Monitors eddy currents to measure proximity, defects, or plate thickness.	Industrial automation

that contains a discrete microprocessor unit. Consequently, there are a large number of instruments that are classified as intelligent sensors. A few examples of intelligent sensors and their applications are listed in Table 11.1. Interested readers are referred to two recent books published in this area [11.7, 11.8].

11.2 Structural Compensation

As stated above, a measure of the intelligence of a sensor is its ability to perform some kind of logical function. For example, the base-line resistance of a pellistor-type gas sensor (§9.6.2) is both sensitive to the ambient temperature and tends to degrade with time. The first effect could be removed perhaps through the separate monitoring of the ambient temperature (e.g. by a thermistor, §5.4) and through the use of an appropriate compensation algorithm. However, the effect, once known, may be greatly reduced by the structural design of the gas sensor. For example, the incorporation of a second pellistor in the reference arm of a Wheatstone bridge (§2.2.1) is an example of an intelligent design which uses structural compensation. This approach can be adopted in a microsensor system with the compensating sensor integrated with the original sensor. However, this type of integrated sensor would not strictly be regarded here as a smart sensor.

When the structural design of a sensor generates more than one signal that needs to be processed this could become the basis of a smart sensor. Taking the example of a thick film resistive gas sensor, the use of two pairs of electrodes, instead of just one, each with a different size of electrode separation provides information on the response from different regions of the film (see Figure 11.2).

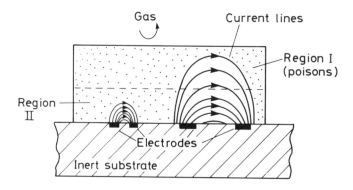

Figure 11.2 Structural compensation by an intelligent thick film gas sensor consisting of two pairs of electrodes.

The ratio of resistances can provide a diagnosis of the relative sensitivity of each region of the active film and so the surface poisoning. Provided that the gas concentration is uniformly distributed throughout the film, then the ratio of the resistance of the wide electrode gap R_w to the narrow electrode gap R_n will be constant until poisoning occurs in the outer region of the active film. This will cause R_w (region I) to be affected before R_n (region II) is. Thus,

$$\frac{d(R_w / R_n)}{dt} \text{ is zero until poisoning occurs.} \qquad (11.1)$$

A self-diagnostic capability such as the above is an important attribute of a smart sensor as discussed later (§11.5).

11.3 Integration of Signal Processing

The integration of the preprocessor, and perhaps the processor itself, onto a single chip can bring many advantages. The most compelling argument for the fabrication of an integrated sensor is to enhance the signal-to-noise ratio. The electrical power output from a microsensor is often low and susceptible to stray capacitance, inductance and noise. In these cases, the direct transmission of the output signal down a long interconnecting wire may be impractical. This is particularly true when, for example, a capacitive pick-up is used in a microsensor. The capacitance of a lateral resonant silicon structure (§7.2) may be only a few femtofarads. Thus, on-chip FET circuitry is highly desirable in order to remove the effect of the high input capacitance of transmission cables and subsequent instrumentation (~ pF). Clearly, this will improve the response, sensitivity and resolution of the microsensor. There are other reasons for integrating the signal processing with the sensor such as providing an enhanced functionality at a lower cost, size or weight.

11.4 Self-calibrating Microsensors

Most sensors have at least two parameters that need to be set during the manufacturing process, the offset and the sensitivity or gain. In the mass-production of a sensor, the process of calibrating individual sensors (§10.2) is expensive and undesirable, but often essential. This cost can re-occur during the life-time of the sensor because of parametric drift. So maintenance work is carried out to recalibrate manually the offset or gain. Consequently, there is considerable need for a self-calibrating sensor which can carry out its own calibration, and this is particularly true when a high level of precision is required.

The conventional calibration of a sensor may involve the laser trimming of integrated resistors. This means that the resistor film and patterning process must be compatible with the IC and microsensor technology. Typically, resistor films (e.g. Ta, Cr, NiCr) are vaporised from the substrate by a YAG laser in controlled cuts or trims. Figure 11.3 shows some common resistor cuts. Trimming at the wafer level may not be possible for some types of smart sensors. In these cases,

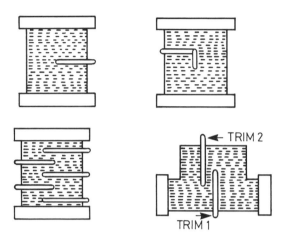

Figure 11.3 Conventional trimming patterns of resistor films.

the analogue IC must compensate for the lower level of process control using, for example, voltage reference signals, or removing the need for a null offset. Clearly, the laser trimming of a resistor film is a one-off process and does not solve the problem of parameter drift. Local recalibration or self-calibration may still be necessary during the life of the sensor.

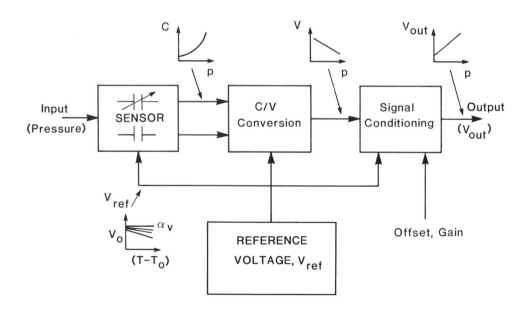

Figure 11.4 Schematic representation of the signal processing in a smart capacitive pressure sensor. Adapted from [11.9].

One example of a reported smart sensor with a self-calibration capability is in pressure sensing (§7.6.4). Figure 11.4 shows a block diagram of the signal processing that takes place in the capacitive sensor and its associated electronic circuitry. The capacitive pressure microsensor has on-chip CMOS switched-capacitor circuitry to perform accurate capacitance-to-voltage conversion, signal conditioning and an automatic compensation for the device temperature using a reference voltage V_{ref} [11.9]. The temperature-sensitivity of this smart sensor α_v is related to a reference voltage V_{ref} by,

$$V_{ref} = V_0\left[1 + \alpha_v\left(T - T_0\right)\right] \qquad\qquad (11.2)$$

and α_v is programmable with a 3 bit resolution.

11.5 Self-testing Microsensors

The ability of a sensor to test its functionality is highly desirable. Recent developments in the field of smart sensors are leading to sensors with some limited diagnostic capability. This is basically an ability of a sensor to determine whether it is functioning normally. A complete failure would usually be detected by the user as the output, either current or voltage, falls below its operating specification. However, in many cases a sensor can fail to perform adequately but provide a reasonable output (§10.4). In these cases more sophisticated quality assurance is necessary. For instance, a noise characteristic (e.g. power spectral density) may be related to the physical property of the sensing element that is changing and affecting its performance and can thus be used to provide diagnostic information.

A significant level of self-testing has been reported on a smart microaccelerometer [11.10] in which the basic mechanical performance can be tested routinely and thus diagnosed for faults.

11.6 Multisensing

Smart sensors often improve their performance through the use of other sensors to monitor undesirable dependent variables. For example, the temperature-sensitivity of a microsensor could be ameliorated via the integration of a thermodiode near the sensing element.

A smart sensor can also eliminate spurious rather than systematic erroneous signals. For instance, an array of identical sensors can be employed and coupled to a microprocessor which calculates the average sensor output or perhaps discards any anomalous readings in a voting logic. The former approach has been adopted to make a smart pH sensor with 10 identical sensing elements [11.11]. The set of sensor signals permits both a higher output voltage, and a greater reliability (§10.1) as the fault-tolerance is improved.

Some recent developments in the field of sensor array devices are discussed in Chapter 12. These are usually integrated sensors which may also have some higher level of processing capability (i.e. intelligence). For example, the use of an array of dissimilar chemical microsensors and an artificial neural network processor has led to the development of an intelligent electronic nose (§12.5.3).

11.7 Communication

The read-out electronics may be accommodated in a smart sensor. The simplest form of the read-out circuitry is often an analogue current or voltage output. For example, the 4 to 20 mA current loop is a common standard and provides some immunity to noise. A second approach is to use frequency modulation. For example, resonant microsensors (§7.2) produce an oscillatory output signal which can be either counted or converted to a voltage by a circuit integrated onto the sensor chip. Moreover the signal can be then be converted to a digital signal on the chip, and the digital output interfaced to a bus system (§2.5). These functions are commonly integrated onto a smart sensor although the reasons for doing so are often scientific rather than hard commercial decisions. One important consideration when deciding whether to fabricate an integrated sensor is the compatibility of the materials' processing required, in particular the temperature range over which the technology bases operate, see Figure 11.5. Bulk CMOS is relatively inexpensive but is limited to a low temperature range and may be unsuitable when some inorganic active layers are required. In contrast, Silicon-On-Insulator (SOI) can withstand a higher processing temperature but is a much more expensive process (e.g. SOS).

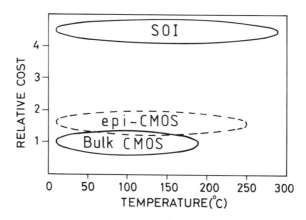

Figure 11.5 Relative costs of smart sensor IC technologies as a function of the processing temperature.

11.8 Applications of Smart Sensors

Currently, there are relatively few applications of smart sensors; however, this situation is likely to change significantly during the next decade. The field of radiation microsensors (Chapter 6) has had some notable successes: cameras with on-chip amplification and temperature compensation are on the market. Smart optical sensors are relatively easy to make because the silicon sensor technology is compatible with the IC processing. CCD image sensors with integrated image intensifiers have also enjoyed some commercial success. Such array sensors are, however, expensive and are the topic of discussion in the next chapter (§12.2).

Smart mechanical sensors, such as pressure sensors (§7.6.4), were initially fabricated at a relatively high cost and in low volumes. However, the interest of the automotive industry in low-cost microaccelerometers for air-bags has changed this situation.

Smart silicon Hall plate devices (§8.2.2) have been fabricated which include a built-in offset for the null voltage and internal temperature compensation.

Perhaps the most difficult, but potentially most rewarding, development in the field of smart sensors will be in chemical and biological sensing (Chapter 9). Traditionally, chemical sensors suffer from various problems, such as drift in any offset parameters, degradation of sensitivity due to poisoning, and interference from humidity and other chemical species. The design and integration of intelligent processing algorithms in (bio)chemical microsensors could correct these deficiencies and would create large markets in fields such as environmental monitoring and medical diagnostics.

Suggested further reading

Readers are referred to the following texts that provide some background information for this chapter on smart sensors:

Brignell J and White N: *Intelligent Sensors* (1994). Published by IOP Publishing Ltd, UK. ISBN 0-7503-0297-6. (272 pages. Excellent discussion of the nature and application of intelligent sensors)

Ohba R (ed.): *Intelligent Sensor Technology* (1992). Published by Wiley & Sons Ltd, UK. ISBN 0-471-93423-2. (167 pages. Overview of intelligent sensor technology as defined by microprocessor-based sensing instrumentation)

References

11.1 Middelhoek S and Hoogerwerf AC (1985) Smart sensors: when and where? *Sensors and Actuators*, **8**, 39-48.

11.2 Hauptmann P (1991) *Sensors: Principles and Applications*, Prentice Hall International (UK) Ltd, UK, p. 7.

11.3 Breckenbridge RA and Husson C (1978) Smart sensors in spacecraft: the impact and trends. *Proc. AIAA/NASA Conf. on Smart Sensors, Hampton, USA*, pp. 1-5.

11.4 Giachino JM (1986) Smart sensors. *Sensors and Actuators*, **10**, 239-249.

11.5 Khanna T (1990) *Foundations of Neural Networks*, Addison-Wesley Publishing Company Inc., USA, 196 pp.

11.6 Forsyth RF (ed.) (1984) *Expert Systems: Principles and Case Studies*, Chapman and Hall, London, 231 pp.

11.7 Zuech N (ed.) (1991) *Handbook of Intelligent Sensors for Industrial Automation*, Addison-Wesley Publishing Company, Inc., USA, 521 pp.

11.8 Ohba R (ed.) (1992) *Intelligent Sensor Technology*, John Wiley & Sons Inc., New York, 167 pp.

11.9 Schnatz FV, Schöneberg U, Brockherde W, Kopystynski P, Mehlhorn T, Obermeier E and Benzel H (1992) Smart CMOS capacitive pressure transducer with on-chip capability. *Sensors and Actuators A*, 34, 77-83.

11.10 Allen HV, Terry SC and de Bruin DW (1989) Accelerometer systems with self-testable features. *Sensors and Actuators*, **20**, 153-161.

11.11 Cheung PW (1982) Recent developments in integrated chemical sensors. *Proc. 1981 Int. Electron Devices Meeting, Washington DC, USA*, pp. 110-113.

Problems

11.1 What is the difference between an integrated sensor and a smart sensor? Give one example of each type of sensor.

11.2 Briefly describe three ways in which a sensor may be regarded as having intelligence.

11.3 State whether you would design the following devices to be smart, giving the reasons for doing so: (a) a lateral resonant capacitive pressure microsensor, and (b) a metal oxide semiconducting chemoresistive microsensor. *(You may wish to refer to other chapters in order to answer this problem)*

11.4 Describe two different fields of application in which you think that smart sensors could play a vital role. Explain why, and give a cost-benefit analysis. *(You may wish to refer to other chapters in order to answer this problem)*

12. Microsensor Array Devices

Objectives

☐ To introduce the topic of microsensor array devices

☐ To present some examples of microsensor array devices

☐ To discuss their application

12.1 Basic Considerations and Definitions

12.1.1 Definitions of array sensors

There is a considerable requirement to monitor and control many of the complex processes which take place in the real world. This has led to the need for sensor systems that are capable of processing signals from a multitude of sources. The general layout of such a system is shown in Figure 12.1. A set of sensors, each

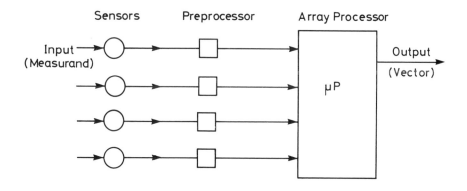

Figure 12.1 General arrangement of an intelligent sensor array system.

with its own preprocessing unit, is connected to an array processor which computes the required output to, perhaps, a display device or an actuator. Each of the sensors in the array can measure the input signal from a different location in space and so can provide the user with spatial information. For example, it may be necessary to measure the concentration of methane gas in air at various points down a mine shaft, or across a waste disposal site. Clearly, in these applications the distance across which the data are gathered requires a distributed array of discrete and identical sensors rather than an integrated array device.

Figure 12.2 shows the general architecture of several types of microsensor array device with various levels of integration as denoted by the dotted lines. When the array of sensors is integrated, we have a Microsensor Array Device (MAD) as part of a sensor array system. Integration of the preprocessor (e.g. amplifier, filter) on a separate device creates a hybrid MAD (HMAD), or onto the MAD creates an integrated MAD (IMAD). Finally, the full integration of the preprocessor and processor with the microsensor array produces a smart MAD (SMAD).

In this chapter, we discuss the various principles, such as radiative, mechanical, magnetic and chemical, and applications of different classes of MADs. However, before doing so, we wish to consider the different types of function that sensor array devices can perform.

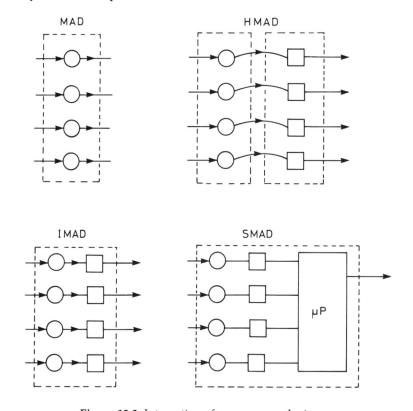

Figure 12.2 Integration of sensor array devices.

12.1.2 *Functionality of array sensors*

The functionality of a MAD is primarily determined by the characteristics of the sensors employed within it. Let us consider a set of n sensors where each sensor i has a response y_{ij} to measurand j. The output from the MAD may be regarded now as a vector y_j of n dimensions where,

$$y_j = \{y_{1j}, y_{2j}, ..., y_{nj}\} \qquad (12.1)$$

If we use a set of identical microsensors in the MAD, then the output vector contains a set of identical scalar components, i.e.

$$y_{1j} = y_{2j} = y_{ij} \qquad (12.2)$$

In other words, each sensor i correlates exactly with each of the other sensors i', so the linear regression coefficients are,

$$\wp_{ii'} = 1 \quad \forall \, i, i' \qquad (12.3)$$

A MAD that contains a set of identical microsensors can be used in a variety of ways. For example, the output of the elements in an array device can be summed to produce an amplification effect with a gain of n,

$$y_j = (y_{1j} + y_{2j} + ... + y_{nj}) \approx ny_{ij} \qquad (12.4)$$

The total output signal y_j is now significantly higher and can thus be processed more readily. A practical example of this is the thermopile (§5.2) which comprises a combination of thermocouples. The voltage difference generated by a single thermocouple is typically 10 μV and so a set of, say, 40 thermocouples would produce a substantial signal of 0.4 mV.

Besides signal amplification, a set of identical sensors can be used to enhance the reliability of the device. For example, a system which comprises two identical sensors, each of which is independent and capable of detecting the measurand, possesses parallel active redundancy (§10.1). In other words, when either of the sensors fails, there is still an acceptable output y_j and so the reliability $\Re_j$ of the sensor system is greater than that for an individual sensor as,

$$\Re_j = \Re_{1j} \oplus \Re_{2j} > \Re_{1j} \qquad (12.5)$$

Extending the sensor array from 2 to n sensors provides a higher level of redundancy, or introducing a logic voting where m out of n sensors are needed to provide an acceptable output. Such a system is said to be fault-tolerant as erroneous signals can be compensated for automatically. The use of redundancy is expensive but common in systems which require a high reliability of performance. Sensors used in safety-critical systems such as nuclear power plant and aircraft would fall into this category. However it is often necessary to separate sensors in order to remove some common modes of failures (e.g. fire) which are potentially hazardous.

Finally, and perhaps more importantly, the signals coming from an array of sensors can be used to form a 1-d or 2-d spatial map of the measurand. A MAD can thus be used to measure the periodic variation of the measurand when the sensors are placed a uniform distance apart, d. The response y_j of a linear array of sensors to measurand j is

$$y_j = \{y_{dj}, y_{2dj}, ..., y_{ndj}\} \tag{12.6}$$

Linear array devices, such as those employed in line-scan cameras (§12.2.2), can profile the intensity of light along a line with time. Using a suitable optical system, a moving object can then be imaged by a line-scan camera.

A set of (n × n) identical sensors generates a two-dimensional response matrix map of the measurand, represented in matrix notation by

$$\tilde{y}_j = \begin{bmatrix} y_{11} & y_{21} & \cdots & y_{n1} \\ y_{12} & y_{22} & & \vdots \\ \vdots & & \ddots & \vdots \\ y_{1n} & y_{2n} & \cdots & y_{nn} \end{bmatrix} \tag{12.7}$$

One advantage of using a 2-d array of sensors is that the output signals from an area or region can be processed concurrently without the need to move either the object being sensed or the sensing device. A common example of this type of array device is the CCD used in video cameras (§12.2.2).

Figure 12.3 summarises some of the different functions that can be obtained from a set of identical sensors. The integration of identical sensors into a microsensor array device is relatively straightforward for silicon sensors. The further integration of the preprocessor is also common. Feature extraction and other pattern recognition techniques can be applied to the output of a MAD with identical sensors (e.g. a CCD array).

It is also possible to employ either completely different or dissimilar (in terms of the measurand) sensors within a sensor system. In a sensor array which comprises completely different sensors, then each sensor is independent and responds to different measurands (e.g. a thermodiode and a MAGFET). The vector response is now simple because only one of the n sensors responds to the measurand j or j′, e.g.

$$y_j = \{y_{1j}, 0, ..., 0\} = y_{1j} \quad \text{and} \quad y_{j'} = \{0, y_{2j'}, ..., 0\} = y_{2j'} \tag{12.8}$$

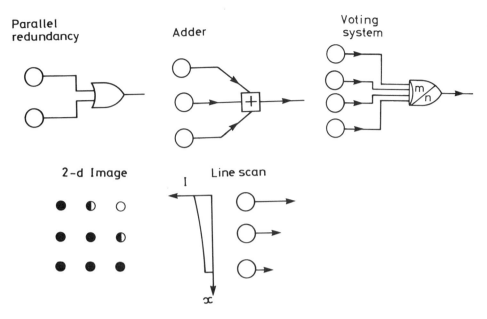

Figure 12.3 Some functions that can be carried out by a set of identical sensors: adding, parallel redundancy, voting logic, line-scanning and imaging.

In other words, the cross-correlation of all terms off the leading diagonal is zero. The principal advantage in constructing a microversion of such an array device is a reduction in size but this may be offset by the cost of combining various sensing technologies onto a single device.

A more interesting situation occurs when more than one of the sensors is sensitive to a measurand - but to a different degree. For example, a pressure sensor (§7.6.4) has a primary role of measuring pressure but its output may also have a dependency upon temperature. So the response of the pressure sensor y_{PT} may be a complex function of pressure and temperature,

$$y_{PT} = F(P,T) \qquad\qquad (12.9)$$

However, the inclusion of a temperature sensor (Chapter 5) in the system (assuming it is independent of pressure) can be used to correct the output of the pressure sensor. Thus, the integration of an additional sensor can improve the performance of another sensor. This process is sometimes referred to as sensor fusion, and enables the specification of a poor sensor to be substantially enhanced. Clearly, it is necessary for the processor to perform some kind of calibration (§10.2) or feature extraction. So this type of MAD has a higher level of intelligence than the previous ones that have been described.

Another type of array may possess a set of sensors which all respond broadly to one class of measurand but have subtle differences (i.e. partially correlated outputs). An example of this is an Electronic Nose (§12.5.3) where the chemical sensors in an array have a poor specificity and so they respond to a very wide

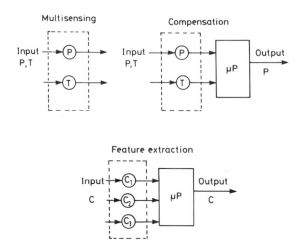

Figure 12.4 Three functions that may be carried out using an array of different and dissimilar sensors.

range of odours. The array processor must now carry out a sophisticated analysis of the response vectors to extract the subtle differences in signal. One possible approach is to use a neural classifier to identify the different measurands (e.g. smells). Figure 12.4 illustrates these three different types of array systems.

Table 12.1 summarises the different categories of MADs discussed here together with a brief description of their features and application areas.

A more detailed discussion is now given on the different types of MADs that have been reported in the literature or are commercially available.

Table 12.1 General characteristics of microsensor array devices.

Sensors	Processor	Special function(s)	Example
Identical	Adder	Signal amplification, noise reduction	Thermopile
	Or logic	Parallel redundancy, enhanced reliability	Process plant
	Voting logic	Fault-tolerance	Safety-critical systems
Different	Multiplexer	Concurrent monitoring of many variables	Multisensing
	Microprocessor	Automated compensation of dependent variable	Pressure sensor with temperature compensation
Similar	Microprocessor	Extraction of features from multivariate data	Electronic Nose, intelligent cameras

12.2 Radiation Array Sensors

12.2.1 Thermal array sensors

As described above, it is possible to use an array of thermocouples to increase the sensitivity of a thermoelectric temperature sensor (§5.2). This concept can be extended to produce a thermopile from a silicon-based sensing array for detecting infra-red radiation. Figure 12.5 shows the arrangement of such a fabricated silicon thermopile detector [12.1]. The chip measures 11 mm by 5.5 mm and consists of a (16 × 2) array of polysilicon/gold thermocouples on top of a 1.3 μm thick SiO_2/Si_3N_4 diaphragm. Table 12.2 details the performance of the elements in this silicon thermopile linear array. The MAD has on-chip multiplexing which is fabricated using a silicon-gate E-D PMOS process. The spacing between the elements is 600 μm. The response time of this device is fast (5 to 10 ms) and it provides a thermal map of a radiated signal.

Since the 1980s considerable progress has been made in the development of infra-red MADs. The most common array sensors being used are extrinsic silicon photoconductors, CdHgTe photovoltaic arrays, Pt-Si Schottky diode arrays and pyroelectric sensors [12.2].

Extrinsic Si photoconductors are made from silicon doped with, for example, Ga to control the band-gap and hence the cut-off wavelength. These devices are operated at very low temperatures (~77 K) but achieve a high level of performance. Infra-red cameras have been developed for satellite imaging such as the ISOCAM camera which consists of a (32 × 32) element Si:Ga array manufactured by CEA-LET-LIR [12.3]. The development of (64 × 64) and (128 × 128) infra-red photoconductor cameras is imminent.

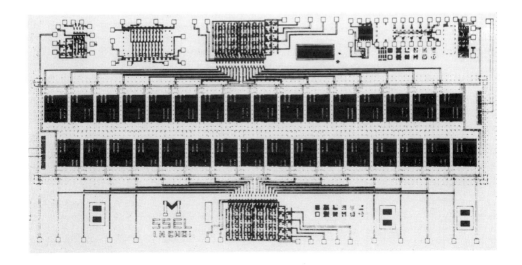

Figure 12.5 Example of an IMAD: a silicon thermopile array IR detector with 32 thermocouples per window [12.1] (© 1986 IEEE).

Table 12.2 Typical performance of thermopile array elements in an IMAD [12.1] (© 1986 IEEE).

Parameter	Measured values	Conditions
Sensing materials	Boron-doped poly-Si/Au	-
Detector area, A	400 μm × 700 μm	-
Diaphragm thickness	1.3 μm	-
Resistance, R	42 to 47 kΩ	25°C, DC
Response, y	8.4 to 12.6 V/W	230°C, DC
Detectivity, D^*	1.6 to 2.5×10^7 cm√Hz/W	230°C, DC, 1 Hz
Time constant, τ	5 to 10 ms	63% of peak to peak

CdHgTe photovoltaic arrays show great promise due to both their wide IR range (1 to 50 μm) and their lower thermal noise generation. IR cameras consisting of an array of (256 × 256) photodiodes have been reported [12.4]. The use of indium bump hybridisation has permitted the combination of CdHgTe with silicon read-out circuits which is particularly attractive, see Figure 12.6.

Although PtSi Schottky diode arrays have a limited spectral response (< 5 μm) and a low quantum efficiency, arrays of high density can be made with highly-developed silicon microprocessing technology. A monolithic array of (512 × 512) elements has been reported [12.5] which is compatible with VLSI technology.

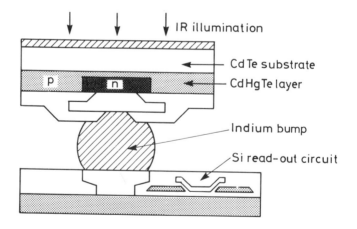

Figure 12.6 Bump bonding of a CdHgTe photodiode array to a silicon read-out circuit [12.2].

Infra-red array devices are also being developed that employ pyroelectric detectors (§6.4.4). These promise a low-cost device with reasonable performance at ambient temperatures.

12.2.2 Optical array sensors

The development of optical linear MADs has been an enormous technical and commercial success. For example, Figure 12.7 shows a photograph of a family of commercial photodiode arrays having DIP-type ceramic cases and quartz glass windows. The typical characteristics of these commercial (Hamamatsu) photodiode arrays are given in Table 12.3 below.

Table 12.3 Some commercial linear photodiode arrays. (Courtesy of Hamamatsu, Japan.)

Number of photodiodes	Spectral range (nm)	Type no.	Features
35, 38, 46	190 to 1,100	S-4111	Wide spectral response, good linearity and fast response (~ 1 μs). Applications in light spectrum analysers, position detectors.
128, 256, 512, 1024	200 to 1,000	S-3900	Wide dynamic range, low power consumption MOS devices, serial/current output.

Figure 12.7 Photograph of the S-3900 family of MOS linear photodiode arrays. (Courtesy of Hamamatsu, Japan.)

The output from a photodiode linear array is usually read in one of two ways, either in real time using a set of op-amps before a multiplexer, or sequentially using only one op-amp after the multiplexer when the charge has been stored in the junction capacitance of each channel, as illustrated in Figure 12.8.   The advantage of the first method is that the signal is conditioned (i.e. low output impedance) before the multiplexer so noise is less of a problem than in the second method.

The output from a photodiode array can be connected to, for example, a single video output line using integrated digital shift registers.   Figure 12.9 shows the

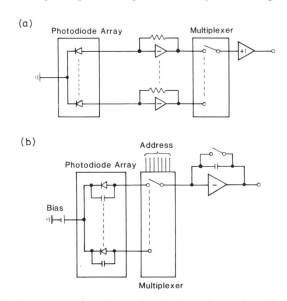

Figure 12.8 Two types of read-out circuits for a linear photodiode array: (a) charge coupled, and (b) charge stored.

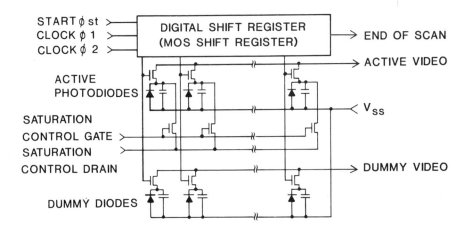

Figure 12.9 Driving circuit for a linear photodiode array (e.g. S-3900 family).

electronic circuit for the read-out of a MOS linear photodiode array.

The output from a photodiode depends upon the intensity of the light falling on it. Various types of read-out circuit have been developed, some of which provide a higher sensitivity by integrating the charge generated and running at lower clock-rates. Thus a wide variety of applications are possible, even when the light levels are low.

It is clearly a short step to extend the fabrication of a linear array to a 2-d array to produce a two-dimensional imaging array. Silicon image sensors use either an n^+-p photodiode or an MOS capacitor. The read-out electronic circuit can employ MOS switches and sense lines for small arrays but the relatively high capacitance of a sense line can cause a significant drop in signal [12.6]. Instead, a charge-couple device (CCD) has a transfer register that permits each row of sensor data to be moved in a serial manner while light is being accumulated in a cell storage array. The CCD is the most popular means of transferring an optical image or frame from the light-sensitive array to an output as it overcomes the line capacitance problem. However, in some applications the delay caused by the serial transfer is unacceptable and so a charge-injected device (CID) can be used. The CID permits the individual accessing of the photosensitive cells although it can be made to scan in a conventional raster pattern. Silicon image sensors offer substantial benefits over a conventional vidicon tube, except on cost and resolution. However, these two disadvantages are likely to disappear in the near future. Table 12.4 makes a brief comparison of these imaging technologies [12.7].

Table 12.4 Relative comparison between imaging devices used in video cameras [12.7]. (Reproduced by permission of Prentice-Hall Inc., ©1989)

Feature/specification	CCD	CID	Vidicon
Relative size	Small	Small	Large
Reliability	High	High	Medium
Resolution	Medium	Medium	High
Sensitivity	Medium	Low	High
Speed	Medium	Fast	Slow
Bloom (error)	Average	Small	Large
Current cost	Medium	High	Low
Future cost	Low	High	Medium

Although cameras that employ CCDs or CIDs tend to be more expensive than vidicon cameras, they have a lower level of geometric distortion (~ 0.1% vs. 1%). Moreover, CID cameras can often be run above standard video clock speeds under suitable illumination, e.g. 20 MHz. The ability of a silicon camera to run at a high speed is particularly important when using them as part of a vision system in a manufacturing process, see Figure 12.10. The video image can be displayed on a monitor and ideally processed on a card connected to the PC bus. Processed images can then be sent down an RS-232C interface (§2.5) when low baud-rates are acceptable.

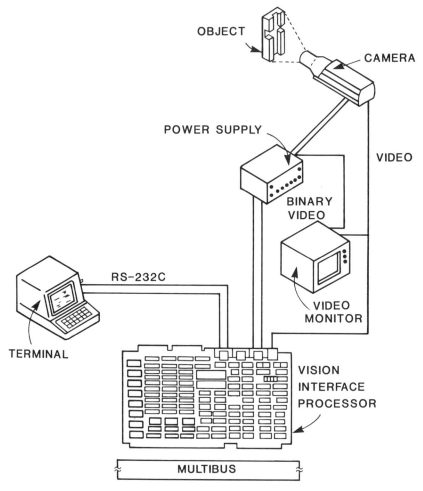

Figure 12.10   Basic layout of an intelligent CCD vision system for analysing objects in an automated manufacturing process.

The automatic inspection of bank-notes for printing defects requires a very fast image analysis system so that the bank-notes can be moved rapidly across the camera's field of view. This requires hard-wired processing circuitry as well as a high speed camera with a substantial increase in cost.

The use of a silicon optical sensor array permits the integration of more sophisticated signal processing circuitry, such as programmable logical devices or digital signal processing chips, for the high speed processing of video signals. These devices can also be used in optical communication.

Another interesting application is in the use of a CCD camera in a robot vision system. Figure 12.11 shows a possible arrangement in which a CCD camera can be used to image an object and provide visual information to a robot. The robot system is then able to detect the position of its end-effector and so control its joint movements accordingly.

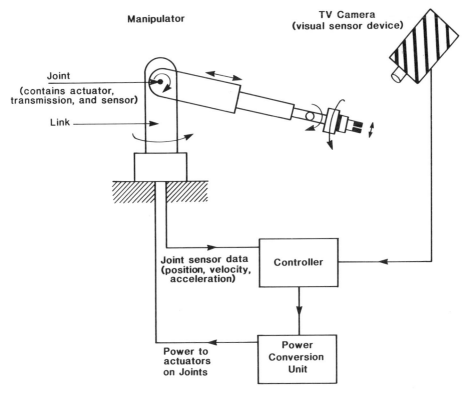

Figure 12.11 A robot with machine vision from a CCD camera.

12.3 Mechanical Array Sensors

The commercial development of mechanical array sensors rather lags behind that of optical arrays, although in a peculiar way it could be argued that a mechanical filter was a mechanical array device which "sensed" the relative size of particles or molecules in a liquid or gas. These "microsensors" are sold in vast numbers. For example, the radiation and chemical etching of a membrane material can produce an irregular array of micron-sized holes in a membrane, see Figure 12.12. The membrane "senses" particles below a certain size by modulating the flux through it. However, mechanical sensor array devices are usually thought of as active rather than passive components.

Mechanical sensors are often used to measure the force or pressure exerted by one object on another (§7.6). The measurement of (simple) touch is of great interest in the field of manufacturing where components are being assembled. The automation of many assembly procedures has led to the widespread use of robot grippers to handle components. Point sensing of the force applied is clearly of importance, but there is an increasing demand for tactile information. The human finger can both detect forces in the range of 0.001 to 0.1 N and discriminate between forces which differ by about 15%. Yet more importantly the finger produces information about the shape and texture of the measurand. The spatial resolution of a finger is about 1 mm and so represents a tactile array of about 15 × 20 elements [12.8]. The development of a tactile sensor array would find application in robotic grippers where such information would allow the sensitive gripping of delicate objects. Other applications would include the inspection of the shape of parts for identification or quality control [12.8].

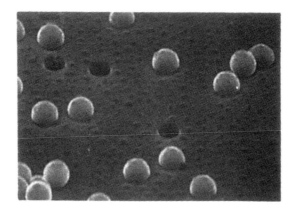

Figure 12.12 High precision filter with 100 nm pore sizes (Bibby Sterilin).

The fabrication of an (8 × 8) silicon-based tactile imager was reported in 1985 [12.9]. The device consisted of an *x-y* arrangement of capacitive force sensors (§7.6) with a maximum operating force of 0.1 N per element. The array was fabricated using standard IC process technology and the read-out electronics employed an MOS switched-capacitor charge integrator. Table 12.5 summarises some of the characteristics of early tactile array sensors.

Table 12.5 Performance of some tactile array sensors [12.9] (© 1985 IEEE).

Device parameter	Size of array: (4 × 4)	(8 × 8)	(16 × 16)
Cell spacing (mm)	4.0	2.0	1.0
Zero-pressure capacitance (fF)	6.48	1.62	0.4
Rupture force (N)	18.9	1.88	0.19
Max. linear capacitance (fF)	4.8	1.2	0.3
Max. output voltage (V)	1.2	0.6	0.3
Max. resolution (bit)	9	8	8
Read-out (access) time (μs)	-	< 20	-

Figure 12.13   Photograph of a 1,024 element pressure MAD with integrated electronic circuitry [12.10].

More recently, a large (32 × 32) element silicon pressure sensor array has been fabricated with CMOS processing circuits [12.10]. The sensor structure consists of polysilicon piezoresistors (§7.6) on top of silicon microdiaphragms integrated using a 3 μm CMOS design process. Figure 12.13 shows a photograph of the entire array [12.10]. The cell access time is only 16 μs so the read-out time of a frame is 16 ms (i.e. ≈ 60 Hz). Although a piezoelectric force sensor is less sensitive and more temperature-dependent than other types of force sensor (e.g. magnetic force-balance), it has the advantage of being readily integrated at low cost. It is likely that we shall see more sophisticated array sensors in the near future which measure complex touch. For instance, the non-normal forces are important to detect shear or slipping movement in robot end-effectors.

Another exciting development in the field of mechanical array devices is in the area of active optical components. Optical lenses and mirrors are made by forming/polishing substrates into spherical (or other) surfaces which can be a time-consuming process. The mechanical grinding of small mirrors is difficult and an entirely unsuitable method for making micromirrors in opto-communication components. Therefore the fabrication of an active mirror using silicon micromachining is highly desirable. Figure 12.14 shows a proposed configuration (IBM Corporation and Texas Instruments Inc., USA) of deformable micromirrors to create an active mirror. Active micromirrors have now been realised at MESA (University of Twente). A micromirror array could be used in many applications, such as the high-speed routing of optical communication signals. The construction of active mirrors also permits the active compensation of any geometric distortion in the substrate caused by processing or temperature variation, and can thus provide an intelligent mirror.

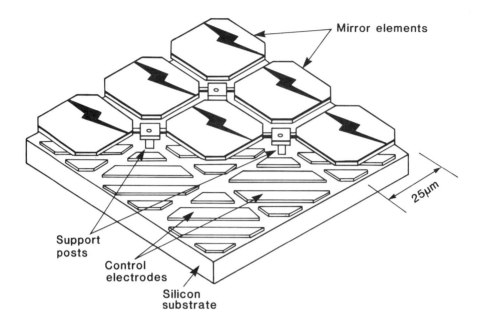

Figure 12.14 An array of active micromirrors [12.11] (© 1989 IEEE).

12.4 Magnetic Array Sensors

The use of an array of magnetic sensors to produce a silicon compass has already been described (§8.2.2). A suitable arrangement of magnetic sensors can readily provide information about the components of magnetic flux density, i.e. B_x, B_y and B_z. More recently, an array CMOS magnetic-field sensor has been reported with current mode output and on-chip reference current sources [12.12]. The device consists of an array of 24 MAGFETs (§8.5) connected in parallel and processed via NMOS and PMOS transistor circuitry, Figure 12.15. The circuit measures the differential current output ΔI at a bias voltage V_g which can be adjusted outside the chip. The array of the magnetic field sensors provides a device with a higher signal-to-noise ratio and greater sensitivity than normal. The device also produces a fairly linear output as shown by Figure 12.16.

12.5 Chemical Array Sensors

12.5.1 Basic principles

Chemical sensors (Chapter 9) form a class of sensor that has generally under-performed when compared with the physical sensors. To begin with they tend to be very sensitive to ambient conditions, such as temperature and humidity. Secondly, chemical sensors often have poor ageing characteristics, that is the chemical sensing element is slowly poisoned and thus changes its response characteristics with time. Thirdly, chemical sensors often have a poor selectivity

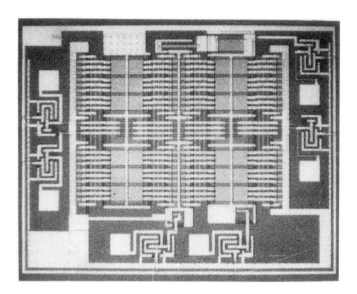

Figure 12.15 A CMOS magnetic-field sensor array with 24 MAGFETs [12.12].

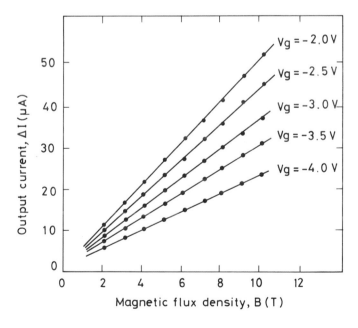

Figure 12.16 Current output of a 24-element MAGFET array sensor at several bias voltages [12.12].

(less so for biosensors) and so interference signals can be generated by other chemical species. For these reasons, researchers have been investigating the use of chemical sensor arrays to improve the overall sensor system performance.

Figure 12.17 shows the basic arrangement of an intelligent array sensor system for detecting chemical species (e.g. gases in air, ions in solutions). The measurand is sensed by an array of sensors, each with its response (e.g. conductance) $G_{ij}(t)$ which is converted to an electrical signal via suitable transduction circuitry. The voltage signals $V_{ij}(t)$ can then be processed to obtain an output parameter for each sensor y_{ij} (e.g. relative change in conductance). When a linear relationship is assumed between the measurand concentration C_j and sensor output y_{ij}, the process can be described using vector-matrix notation by,

$$y = \tilde{A}C + e \qquad (12.10)$$

where y is a row vector of responses with n elements, $\tilde{A}$ is an (n × n) matrix whose elements are the linear regression coefficients (loadings), C is the concentration vector and e is a vector of errors due to the approximation.

For component-specific sensors, the regression coefficient is non-zero only for the measurand and so the matrix $\tilde{A}$ is zero everywhere except on the leading diagonal. In this case there is no need to use any pattern recognition as there is a

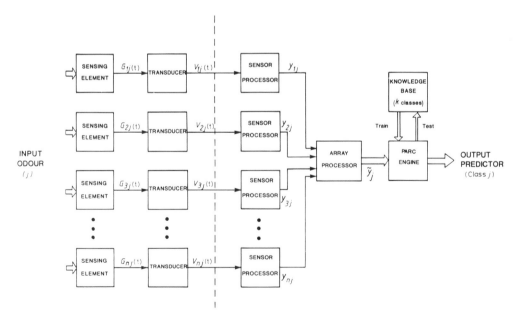

Figure 12.17 Schematic arrangement of an intelligent chemical sensor array system (PARC denotes pattern recognition).

unique solution to the problem. However, in practice some of the off-diagonal terms are non-zero and so array processing is required. Various normalisation procedures and pattern recognition methods have been reported in the fields of gas and odour sensing including non-parametric methods such as artificial neural networks. For a review see [12.13].

12.5.2 Gas sensing

The use of gas array sensors can, in some circumstances, overcome the problem of poor specificity. Various types of gas sensor have been placed in arrays to try and improve their overall performance, see Table 12.6.

Table 12.6 Various types of gas sensors used in arrays.

Sensing element	Elements	Transduction type	Target gases	Ref.
Sintered metal oxide	12	Conductometric	CO, CH_4, H_2	12.14
Phthalocyanines	5	Conductometric	NH_3, H_2S, NO	12.15
Membranes	6	Acoustic (Piezoelectric/SAW)	CO, CO_2, H_2, H_2O, HCl	12.16
Pd-gate MOSFET	10	Threshold voltage	Combustible gases	12.17
Electrochemical cell	2-18	Potentiometric	NH_3, CO, SO_2	12.18

Some success has been reported in the use of parametric techniques to measure the components in gas mixtures with non-specific array sensors [12.13]. However, this is generally limited to simple gas mixtures where the components do not react and thus obey the principle of superposition. Under these conditions, equation (12.10) can be solved using, for example, multiple linear regression or partial least squares techniques [12.19]. More commonly the relationship between the component concentration C_j and sensor output y_{ij} is non-linear. Although non-linear parametric methods can be used, the application of artificial neural techniques is becoming increasingly popular [12.20]. These techniques are based upon a processing architecture that is analogous to the way our human brain works and they are less sensitive to sensor noise and drift than conventional (parametric) methods. No commercial chemical array of this type is yet available, unlike in the related field of odour sensing (see next section).

12.5.3 Odour sensing

At this point in time silicon microsensors have been developed that are capable of mimicking the human sense of sight (e.g. a CCD), touch (e.g. a tactile sensor array) and hearing (e.g. a silicon microphone) but what about the more ephemeral sense of smell? The human sense of smell is still the primary method for the evaluation and control of odours and flavours in a wide range of industrial sectors. These include the food industry, the beverage industry and the perfume industry. In all of these industries there are significant potential benefits in terms of improved quality control, process control and product design from the introduction of gas sensors.   By comparison the existing chemical techniques, such as gas chromatography and mass spectrometry, are complex, expensive and often lack the sensitivity required.

The human olfactory system comprises an array of some 100 million receptor cells located in the olfactory epithelium in the nasal cavity.   These biological sensors are believed to have a poor specificity and low sensitivity rather like electronic chemical sensors.   These primary cells respond to odorant molecules which are typically small polar molecules with a molecule weight between 17 and 200 Da.   The human nose significantly improves its sensor performance by having several layers of neural processing which result in a sensitivity at the sub-ppm level, in addition to an ability to discriminate between several thousand odours.

Researchers have been investigating the use of new gas sensor technologies, rather than conventional gas chromatography etc., to produce an electronic olfactory system or electronic nose [12.21].   Table 12.7 summarises some examples of the types of gas sensors that are used to detect odorant materials, such as coffees and beers, together with their basic sensing principle and the number of elements in the sensor array.

Table 12.7 Examples of odour array sensors.

Sensing element	Elements	Principle	Application
Sintered metal oxide	12	Conductometric	Discrimination of coffee blends and roasts [12.22]
Conducting polymers	12	Conductometric	Detection of off-flavours in lagers and beers [12.23]
Lipid coatings	6	Piezoelectric crystal balance	Discrimination of alcoholic drinks [12.24]
Metal FET structure	-	Photocapacitive current	Production of odour images for alcohols, ammonia, etc. [12.25]
Electrochemical cell	16	Amperometric	Identification of toxic vapours [12.26] and cereals

A common application of an odour array sensor is to discriminate between different odour types or notes.   This differs from the gas array sensor in that it is necessary to map the response from the sensor array onto fuzzy organoleptic (smell) data rather than crisp gas concentration data.   The organoleptic data are needed as a training set to teach the electronic nose the different smells experienced by humans.

The main pattern recognition methods that have been applied in electronic nose technology have been reviewed elsewhere [12.13, 12.21] but the common ones are listed below in Table 12.8, together with the assumptions made about the nature of the data.

Table 12.8 Common pattern recognition methods used with odour array sensors.

Method	Type	Comment
Template matching	Parametric	Comparison of response vectors in a t-test statistic. Simple but prone to sensor drift.
Principal component analysis	Non-parametric	Linear method to display differences.
Cluster analysis	Non-parametric	Linear (Euclidean) method to display differences.
Discriminant function analysis	Parametric	Requires multinormal data but can predict class membership.
Back-propagation (ANN)	Non-parametric	An excellent neural predictive classifier for array signals.

Figure 12.18(a) shows the discrimination of roasted Brazilian and Colombian coffee beans using a 12-element conducting polymer nose [12.27]. The response vector y contains the fractional change in conductance of each sensor, hence the scalar components are

$$y_{ij} = \frac{\left(G_{i,odour} - G_{i,air}\right)}{G_{i,air}} \qquad (12.11)$$

The response vectors were then analysed using principal component analysis to find the set of orthogonal linear vectors that best describe the variance. A plot of the scores of the first two principal components showed a clear difference between Brazilian and Colombian coffees.

A commercial electronic nose is now available and is known as the Fox 2000 or Intelligent Nose (Alpha MOS, France). The instrument is based upon an array of six sintered metal oxide gas sensors which respond to a wide range of odorants. The array signals are processed using an artificial neural network (ANN) technique. First, the user trains the electronic nose on known odours and then the artificial neural network can predict the nature of unknown odours, see Figure 12.18(b), with a high success rate.

12.5.4 Concluding remarks

The potential applications of chemical MADs are enormous and range from the monitoring of environmental pollution to the diagnosis of medical complaints (e.g. diabetes). The use of microtechnology permits the fabrication of low-cost microsensor arrays. However, the greatest challenge lies in the development of intelligent signal processing procedures which can overcome the weaknesses of chemical sensors. The further exploitation of artificial neural networking (ANN) techniques looks particularly promising. Moreover, there are substantial benefits in the use of ANNs which can adapt to time-varying signals but still retain their predictive power. It is not surprising that many of the advances in sensor

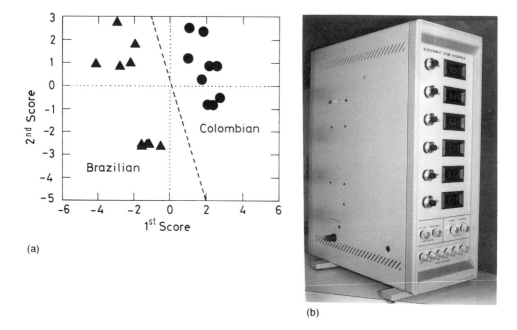

(a)

(b)

Figure 12.18 (a) Discrimination of Brazilian and Colombian coffee beans by a polymer-based electronic nose, and (b) a commercial electronic nose, FOX 2,000 (Alpha MOS, France).

processor technology bring us closer to a system that has been evolving for thousands of years - the human brain. So far the use of biological materials as the active element in sensors and array sensors has been relatively unexplored. Yet it may be possible in the future not only to create complex miniature biological sensors but also to wire them directly into the human nervous system!

Suggested further reading

Readers are referred to the following texts that provide some background information on the application of microsensor array devices (MADs):

Pugh A (ed.): *Robot Sensors* (1986). Published by IFS (Publications) Ltd, UK. ISBN 0-948507-01/2. (315 pages. Basic discussion of the use of optical sensor array devices in volume 1 and tactile and other non-vision sensors in volume 2)

Gardner JW and Bartlett PN (eds.): *Sensors and Sensory Systems for an Electronic Nose* (1992). Published by Kluwer Academic Publishers, Dordrecht, Netherlands. ISBN 0-7923-1693-2. (327 pages. Review of current array technology used for the sensing of odours and gases)

References

12.1 Choi IH and Wise KD (1986) A silicon-thermopile-based infra-red sensing array for use in automated manufacturing. *IEEE Trans. on Electron Devices*, **33**, 72-79.

12.2 Lucas C (1991) Infra-red detection, some recent developments and future trends. *Sensors and Actuators A*, **25-27**, 147-154.

12.3 Lucas C, Mottier P, Ouvrier-Buffet JL and Agnese P (1990) The long wavelength channel detector for the ISOCAM camera. *4th Int. Conf. on Advanced Infra-red Detectors and Systems (IEE Conf. Publ. No. 321), June 1990, London*, pp. 4-8.

12.4 Tennant WE, Kozlowski LJ, Bubulac LO, Gertner ER and Vural K (1990) Advanced materials techniques for large HgCdTe focal plane arrays. Advanced Infra-red Detectors and Systems. *4th Int. Conf. on Advanced Infra-red Detectors and Systems (IEE Conf. Publ. No. 321), June 1990 London*, pp. 15-19.

12.5 Kimata M, Denda M, Yutani N, Iwade S and Tsubouchi N (1987) A 512 × 512 element PtSi Schottky-barrier infrared image sensor. *IEEE J. Solid-state Circuits*, **22**, 1124-9.

12.6 Collett MG (1986) Solid-state image sensors. *Sensors and Actuators*, **10**, 287-302.

12.7 Klafter RD, Chmielewski TA and Negin M (1989) *Robotic Engineering: An Integrated Approach*, Prentice-Hall International Inc., New Jersey, p. 448.

12.8 King AA and White RM (1985) Tactile sensing array based on forming and detecting an optical image. *Sensors and Actuators*, **8**, 49-63.

12.9 Chun K and Wise KD (1985) A high-performance silicon tactile imager based on a capacitive cell. *IEEE Trans. on Electron Devices*, **32**, 1196-1201.

12.10 Sugiyama S, Kawahata K, Yoneda M and Igarashi I (1990) Tactile image detection using a 1k-element silicon pressure sensor array. *Sensors and Actuators A*, **21-23**, 397-400.

12.11 Item in *IEEE Institute, 22 November 1989.*

12.12 Zheng X, Zhang D, Liu L and Li Z (1993) An array CMOS magnetic-field sensor. *Sensors and Actuators A*, **35**, 209-212.

12.13 Gardner JW and Bartlett PN (1991) in *Techniques and Mechanisms in Gas Sensing* (eds. P.T. Moseley, J.O.W. Norris and D.E. Williams), Adam Hilger, Bristol, UK, Ch. 14, pp. 347-381.

12.14 Shurmer HV and Gardner JW (1990) Intelligent vapour discrimination using a composite 12-element sensor array. *Sensors and Actuators B*, **1**, 256-260.

12.15 Cranny AWJ and Atkinson JK (1992) The use of pattern recognition techniques applied to signals generated by a multielement gas sensor array as means of compensating for poor individual element response, in *Sensors and Sensory Systems for an Electronic Nose* (eds. J.W. Gardner and P.N. Bartlett), Kluwer, Dordrecht, Netherlands, 1992, pp. 197-215.

12.16 Ohnishi M, Ishibashi T, Kijima Y, Ishimoto C and Seto J (1992) A molecular recognition system for odorants incorporating biomimetic gas-sensitive devices using Langmuir-Blodgett films. *Sensors Mater.*, **4**, 53-60.

12.17 Sundgren H, Winquist F, Lukkari I and Lundström I (1991) Artificial neural networks and gas sensor arrays: quantification of individual components in a gas mixture. *Meas. Sci. Technol.*, **2**, 464-469.

12.18 Stetter JR (1991) Detection and analysis of foreign odours in grain. *Report to US Dept. of Agriculture, 1991, Contract 90-33610-5088.*

12.19 Manly BFJ (1986) *Multivariate Statistical Methods: A Primer*, Chapman and Hall, London, 159 pp.

12.20 Rumelhart DE and McClelland JL (1986) *Parallel Distributed Processing*, MIT Press, Cambridge, USA, 547 pp.

12.21 Gardner JW and Bartlett PN (eds.) (1992) *Sensors and Sensory Systems for an Electronic Nose*, Kluwer Academic Publishers, Dordrecht, Netherlands, 327 pp.

12.22 Gardner JW, Shurmer HV and Tan TT (1992) Application of an electronic nose to the discrimination of coffees. *Sensors and Actuators B*, **6**, 71-75.

12.23 Pearce TC, Gardner JW, Friel S, Bartlett PN and Blair N (1993) An electronic nose for monitoring the flavour of beers. *Analyst*, **118**, 371-377.

12.24 Ema K, Yokoyama M, Nakamoto T and Moriizumi T (1989) Odour-sensing system using a quartz-resonator sensor array and neural-network pattern recognition. *Sensors and Actuators*, **18**, 291-6.

12.25 Lundström I, Erlandsson R, Frykman U, Hedborg E, Spetz A, Sundgren H, Welin S and Winquist F (1991) Artificial olfactory images from a chemical sensor using a pulse-light technique. *Nature*, **352**, 47-50.

12.26 Stetter JR, Jurs PC and Rose SL (1986) Detection of hazardous gases and vapours: pattern recognition analysis of data from an electrochemical array. *Anal. Chem.*, **58**, 860-6.

12.27 Gardner JW and Bartlett PN (1993) Intelligent ChemSADs for artificial odour-sensing of beers and lagers. *Proceedings of 11th International Symposium on Olfaction and Taste, Sapporo, Japan, July 12-16.*

Problems

12.1 Describe four ways in which an array of sensors can be used to improve upon the performance of a single sensor.

12.2 What type of optical array sensor would you use in the following applications: (a) the inspection of a prototype PCB for wiring faults, and (b) the inspection of a postage stamp for flaws in a high-speed lithographic printing process? Give the reasons for your choice in each case.

12.3 Briefly state the general difficulties in fabricating an array sensor for the measurement of (a) tactile, and (b) chemical signals.

12.4 Estimate the number and density of sensing elements required to make a direct silicon analogue of the following biological functions: (a) touch by a human finger, (b) sight by the human eye, (c) hearing from a human ear, (d) smell by the human nose, and (e) neurones in the human brain. Discuss the feasibility of manufacturing a robot with the sensing capabilities of a human being based on your calculations.

Appendix A.   List of Symbols

Symbol	Description	Units[1]
	Roman letters:	
a_i	i th coefficient	-
a	Acceleration	$m\ s^{-2}$
a	Radius	m
A	The gain of a system, i.e. output/input	-
A	Area	m^2
A_Z	Atomic mass number (in units of Daltons)	Da
b	Base (region or electrode of a transistor)	-
b_m	Mechanical damping coefficient	$N\ m^{-1}\ s$
B	Magnetic flux density or magnetic inductance	$Wb\ m^{-2}$ or T
c	Collector (region or electrode of a transistor)	-
c	Specific heat capacity	$J\ K^{-1}\ kg^{-1}$
c_p	Specific heat capacity at constant pressure	$J\ K^{-1}\ kg^{-1}$
C	Electrical capacitance	F
C_p	Heat capacity at constant pressure	$J\ K^{-1}$
d	Deuterium (particle)	-
d	Thickness, separation or diameter	m
D	Diffusion coefficient	$m^2\ s^{-1}$
D^*	Detectivity of a radiation sensor	$m\ W^{-1}\ Hz^{-0.5}$
e	Emitter (region or electrode of a transistor)	-
e	Electron charge	C
e	A parameter which represents the error in a signal	-
E	Energy	J
E_F	Fermi level	eV
E_g	Electronic band-gap energy	eV
E_h	Enthalpy of a chemical reaction	J
E_H	Hall electric field	$V\ m^{-1}$
E_k	Kinetic energy	J
E_m	Young's modulus	$N\ m^{-2}$ or Pa
E_R	Energy of radiation (particle)	eV
f	Focal distance	m

---

[1] Expressed in terms of SI derived units rather than SI base units given in Appendix C.

Symbol	Description (continued)	Units
f	Frequency	Hz
F	Force	N
$F()$	A mathematical function	-
G	Electrical conductance	S
G_f	Fluidic conductance	$kg\ s^{-1}\ Pa^{-1}$
$G(s)$	System or plant element in a control circuit	-
h	Height	m
h	Planck's constant	J s
$\hbar$	Planck's constant divided by 2π	J s
H	Magnetic field strength	$A\ m^{-1}$
$H(s)$	Feedback element in a control circuit	-
i or I	Electrical current	A
I	Moment of inertia	$kg\ m^{-2}$
I_m	Impulse (mechanical)	N s
I_R	Intensity of radiation (particles)	$W\ sr^{-1}$
J	Current density	$A\ m^{-2}$
k	A constant	-
k	Boltzmann's constant	$J\ K^{-1}$
k_b, k_f	Backward and forward reaction-rate constants	-
k_g	A geometrical factor	-
k_m	Mechanical spring constant	$N\ m^{-1}$
K	Scale factor in dimensional analysis	-
K_{gf}	Gauge factor	-
$K(s)$	Controller element in a control circuit	-
l, L	Length	m
L	Electrical inductance	H
L	Magnetic loss	-
m	A dimensionless constant	-
m	Mass	kg
M	Mutual inductance	H
n	Neutron (particle)	-
n	Electron concentration or density	m^{-3}
N	A number, e.g. the number of turns in a solenoid	-
N_a	Numerical aperture	-
$N(E)$	Density of states at energy E	m^{-3}

Symbol	Description (continued)	Units
N_t	Number flux density	$m^{-3} s^{-1}$
N_Z	Atomic density, e.g. the number of atoms per m^3	m^{-3}
p	Proton (particle)	-
p	Hole concentration or density	m^{-3}
P	Pressure	Pa
P	Thermopower or Seebeck coefficient	$V K^{-1}$
P	Power	W
$\wp$	Sample linear regression coefficient	-
q	Electrical charge	C
Q	Heat energy	J
Q	Resonance factor	-
Q_m	Mass flow-rate	$kg s^{-1}$
Q_v	Volumetric flow-rate	$m^3 s^{-1}$
r_n, r_p	Correction factor to Hall coefficient for electrons and holes	-
$r(s)$	Reference input to servo system in a control circuit	-
R	Electrical resistance	Ω
R_H	Hall coefficient	$m^3 C^{-1}$
R_L	Load resistance	Ω
R_L	Reluctance	H^{-1}
R_q	Surface roughness	m
R_T	Thermal resistance	$W K^{-1}$
$\Re$	The reliability of a microsensor	-
s	Semiconductor thermocouple exponent	-
s	Laplace parameter (a complex variable)	-
S	Sensitivity of a sensor (i.e. the change in the output signal divided by the change in the input signal)	-
t	Time or age	s
t_0	Zero time in a Weibull analysis	s
t_{90}	The time taken for the output signal from a sensor to reach 90% of its final level of response to a measurand	s
T	Temperature	K
T_{bp}	Boiling point temperature	K
T_{mp}	Melting point temperature	K
$u(s)$	Disturbance input (Laplace domain) due to friction, noise, load etc. in a control circuit	-

Symbol	Description (continued)	Units
$v(t)$	Time-dependent voltage signal (also $V(t)$)	V
v_c	The speed of light in a vacuum	m s^{-1}
V	Voltage	V
V_H	Hall voltage	V
$V(s)$	Command input signal to system element in a control circuit	-
w	Width	m
x	Input signal to sensor, distance or displacement	m
$\dot{x}$	Velocity	m s^{-1}
$\ddot{x}$	Acceleration	m s^{-2}
$x(t)$	Input (time-dependent) signal to a sensor, i.e. measurand	-
$X(s)$	Command input signal (Laplace domain) to a control system	-
y_0	Base-line signal	-
Δy	Absolute response from a sensor	-
y_{max}	Maximum output from a sensor	-
Y	Sputter yield rate	m^{-2} s^{-1} A^{-1}
Y	Yield strength of a material	N m^{-2} or Pa
$Y(s)$	Output signal (Laplace domain) from a control system	-
Z	Figure of merit	-
Z_A	Atomic number (number of protons in nucleus)	-
Z^+	Set of all positive integers	-
Greek letters:		
α	Adsorption coefficient of radiation	m^{-1}
α_l	Linear temperature coefficient of expansion	K^{-1}
α_r	Linear temperature coefficient of resistance	K^{-1}
α_θ	Angular acceleration	rad s^{-2}
β	Material constant of a thermistor	K^{-2}
β^+	Positron (particle)	-
β_w	Shape parameter in a Weibull analysis	-
Γ	Band energy level	eV
$\delta(t)$	Impulse function centred on $t=0$, i.e. the Dirac delta function	s^{-1}
δ	A small difference in a parameter	-
Δ	A finite difference in a parameter	-
ε	Electrical permittivity	F m^{-1}
ε	Error in a signal	-
ε_m	Mechanical strain	-

Symbol	Description (continued)	Units
ε_r	Relative permittivity of a dielectric material	-
E	Etch-rate of a material	-
ζ	Mechanical damping factor	-
η	Quantum efficiency	-
η_d	Ideality factor of a diode	-
η_m	Viscosity	$N\ s\ m^{-2}$
η_w	Characteristic life parameter in a Weibull analysis	s
θ	Angle	$^{\circ}$
θ_H	Hall angle	$^{\circ}$
κ	Thermal conductivity	$W\ m^{-1}\ K^{-1}$
K	Kaon (particle)	-
λ	Wavelength	m
λ_f	Failure-rate	s^{-1}
Λ_f	Acoustic constant	$Hz\ m^{-1}$
μ^B	Magnetic permeability of a material	$H\ m^{-1}$
μ_0^B	Magnetic permeability of free space	$H\ m^{-1}$
μ_r^B	Relative permeability of a material	-
μ_n, μ_p	Electron mobility and hole mobility	$m^2\ V^{-1}\ s^{-1}$
$\mu(t)$	Unit step function changing at time $t=0$	s^{-1}
ν	Poisson's ratio	-
ξ	Population standard deviation	-
Ξ_0	Piezoelectric coefficient	$C\ N^{-1}$
Π	Piezoresistive coefficient	$N^{-1}\ m^2$
Π	Reliability factor	-
ρ	Electrical resistivity	$\Omega\ m$
ρ_m	Density (mass)	$kg\ m^{-3}$
P	Partition coefficient of an active material	-
σ	Electrical conductivity	$S\ m^{-1}$
σ_m	Mechanical stress	$N\ m^{-2}$
τ	Time constant	s
υ	Recombination rate	Hz
ϕ	Potential or electromotive force	V
ϕ_B	Potential barrier or work function	eV
Φ	Phase shift	rad
Φ	Magnetic flux	Wb

Symbol	Description (continued)	Units
Φ_R	Flux of radiation (particles)	$m^{-2}\,s^{-1}$
X	Band energy level	eV
ω	Angular frequency	$rad\,s^{-1}$
ω_0	Natural (angular) frequency	$rad\,s^{-1}$
Ω	Solid angle	sr

Mathematical symbols:

Symbol	Description	Units
∇	3-d vector gradient operator	m^{-1}
$\therefore$	Mathematical symbol meaning "therefore"	-
$\in$	Mathematical symbol meaning "is an element of"	-
$\approx$	Mathematical symbol meaning "approximately"	-
$\sim$	Mathematical symbol meaning "of the order of"	-
$\oplus$	Mathematical symbol meaning "logical or"	-
$\otimes$	Mathematical symbol meaning "vector cross product"	-
$\leq$	Mathematical symbol meaning "less than or equal to"	-
$\geq$	Mathematical symbol meaning "greater than or equal to"	-
$<<$	Mathematical symbol meaning "much less than"	-
$>>$	Mathematical symbol meaning "much greater than"	-
$\tilde{A}$	Notation for a matrix (vectors are in bold typeface)	-
$\bar{x}$	Average value of parameter, x	-
$\equiv$	Mathematical symbol meaning "is equivalent to"	-
$\propto$	Mathematical symbol meaning "proportional to"	-
∞	Mathematical symbol meaning "infinity"	-
$\forall$	Mathematical symbol meaning "for all"	-

Appendix B.   Selected Definitions and Acronyms

Term[2]	Definition
AC	Alternating Current.
Active Sensor	See *Modulating Sensor*.
Actuator	A type of transducer that converts an electrical signal into a non-electrical quantity (e.g. a stepper motor or an electric pump).
AFM	Atomic Force Microscopy.
ANSI	American National Standards Institute.
APD	Avalanche Photo-Diode.
ASIC	Application Specific Integrated Circuit.
Availability	The probability that a device is able to carry out its desired function when required to do so.
Band-gap	The energy separation between the top of the valence band and the bottom of the conduction band.
Band-width	The dynamic (frequency) operating range of a device.
Baryon	A heavy nuclear particle, e.g. a proton.
Base-line	Output signal of a sensor when there is no measurand, i.e. zero point.
Baud	A speed of signalling, one operation per second or 1 bit/s in binary coding.
bp	Boiling point.
BS	British Standard.
Bus	An electrical conductor having low impedance to which two or more devices can be separately connected. See *GPIB*.
ca.	Approximately.
Capability	The capacity of a device to perform its desired function under stated conditions.
cf	Compare with.
CMOS	A logic family based on Complementary Metal Oxide Semiconductor technology.
CMRR	Common Mode Rejection Ratio.
Converter	Any device that carries out simple signal processing operations, e.g. amplification, filtering and analogue to digital conversion.
CVD	Chemical Vapour Deposition.
DAMS	Differential Amplification Magnetic Sensor.

---

[2] Terms which first appear in the main text in italic font are defined here.

Term	Definition (continued)
DC	Direct Current.
d.i.l.	Dual in line (electronic package).
ECE	Electro-Chemical Etch-stop.
EDM	Electro-Discharge Machining.
EDP	Ethylene Diamine Pyrocatechol (etchant solution).
Electric Transducer	See *Processor*.
e.m.f.	Electromotive force.
EPROM	Erasable Programmable Read Only Memory.
FET	Field Effect Transistor.
FIBM	Focused Ion Beam Milling.
Fluor	Fluorescein, a fluorescent dye.
FSD	Full Scale Deflection.
FSO	Full Scale of Operation.
Gain	Ratio of output signal to input signal.
GGG	Gadolinium Gallium Garnet.
GPIB	General Purpose Instrumentation Bus. *Aka* IEEE-488.
Gravimetry	The measurement of weights.
HEL	Higher Explosive Limit.
HMAD	Hybrid Microsensor Array Device.
Hygroscopic	Having a tendency to absorb moisture.
Hysteresis	Difference in sensor output when a measurand is increasing and decreasing.
IC	Integrated Circuit.
ICC	Intelligent Crate Controller.
IEE	Institution of Electrical Engineers (UK).
IEEE	Institute of Electrical and Electronic Engineers (USA).
IMAD	Integrated Microsensor Array Device.
Input Transducer	See *Sensor*.
Integrated Sensor	A sensor integrated with part of the processor, i.e. its preprocessor or modifier.
Intelligent Sensor	A sensor which has one of the following features: a microprocessor unit using mathematical operations, compensating algorithm or 2-way communication.
Interface	A processing unit which lies between the sensor and the processor or some other device.
IR	Infra-Red (radiation).

Term	Definition (continued)
ISFET	Ion Selective Field Effect Transistor.
LCD	Liquid Crystal Display.
LED	Light Emitting Diode.
LEL	Lower Explosive Limit.
Leptons	Light nuclear particles, e.g. an electron.
LPCVD	Low Pressure Chemical Vapour Deposition.
LR	Loading Roughness.
LUT	Look Up Table.
LVDT	Linear Variable Differential Transformer.
MAD	Microsensor Array Device.
Magnistor	A magnetotransistor.
MCT	Mercury Cadmium Tellurium (radiation sensing material).
Measurand	The physical or chemical quantity to be measured (e.g. displacement, pressure or gas concentration).
Mesa	A term used to describe a high, steep-sided plateau microstructure.
Meson	A medium weight nuclear particle, e.g. a kapton.
Microsensor	A miniature sensor that has a sensing element with at least one physical dimension at the sub-millimetre level (i.e. less than 100 microns).
Modulating Sensor	A sensing device that requires the use of an external power supply (e.g. a photocell or thermistor).
MOSFET	Metal Oxide Field Effect Transistor.
mp	Melting point.
NEP	Noise Equivalent Power.
NTC	Negative Temperature Coefficient.
Off-set	Systematic error in the output of a sensor.
Output Transducer	See *Actuator*.
Passive Sensor	See *Self-generating Sensor*.
Pc	Phthalocyanine (a chemically active organic material).
PC	Personal Computer.
PD	Photo-Diode.
Pellistor	A type of calorimetric gas sensor.
PID	Proportional Integral and Differential (3 term) controller.
Preprocessor	See *Converter*.
Processor	Any device that modifies the electrical signal from a sensor (n.b. often a microprocessor when a preprocessor is specified).

Term	Definition (continued)
PSG	Phospho-Silicate Glass (a sacrificial layer in micromachining).
PSR	Power Stress Ratio.
PTAT	Proportional To Absolute Temperature.
PTC	Positive Temperature Coefficient.
PZT	A material used in acoustic sensors.
QCM	Quartz Crystal Microbalance.
RAM	Random Access Memory (electronic device).
Range	The span of the output signal of the sensor, e.g. 0 to 5 V DC.
Reactance	The part of the total impedance of a circuit not due to resistance.
Reliability	The ability of an item to function under stated conditions for a stated period of time (see BS 4778 Part 1: 1979 on Quality vocabulary).
Reluctance	The ratio of the magnetomotive force to the total magnetic flux.
Resist	A photosensitive material used in micromachining.
R.F.	Radio Frequency.
R.H.	Relative Humidity.
r.m.s.	Root mean square.
SAW	Surface Acoustic Wave.
SC	Single Crystal (e.g. a silicon wafer).
Self-exciting Sensor	See *Self-generating Sensor*.
Self-generating Sensor	A sensing device that does not require the use of an external power supply (e.g. a thermocouple or a fuel cell).
SEM	Scanning Electron Microscopy.
Sensor	Any device that converts a non-electrical physical or chemical quantity (e.g. pressure, temperature or humidity) into an electrical signal.
SF	Safety-Factor.
SIBM	Showered Ion Beam Milling.
SM	Safety-Margin.
Smart Sensor	An integrated intelligent sensor, i.e. a sensor with its processing unit integrated onto a single silicon chip.
SOI	Silicon-On-Insulator (processing technology).
SOS	Silicon-On-Sapphire (processing technology).
SQUID	Superconducting Quantum Interference Device.
STM	Scanning Tunnelling Microscopy.
TCR	Temperature Coefficient of Resistance

Term	Definition (continued)
Thermistor	A temperature-dependent resistor.
TI	Texas Instruments, a US company.
Transducer	Any device that converts a non-electrical physical or chemical quantity (e.g. pressure, temperature or humidity) into an electrical signal <u>or vice versa</u>.
TTL	A logic family using Transistor-Transistor Logic.
UV	Ultra-Violet (radiation).
VDU	Visual Display Unit (e.g. a computer monitor).

Appendix C.   List of Base SI Units

Quantity	Unit	Abbreviation
Fundamental:		
Mass	kilogram	kg
Length	metre	m
Time	second	s
Temperature	kelvin	K
Electrical current	ampere	A
Luminescent density	candela	cd
Amount of substance	mole	mol
Supplementary:		
Plane angle	radian	rad
Solid angle	steradian	sr

Appendix D. Unit Prefixes for SI Units

Prefix	Symbol	Multiply by
yacto	y	10^{-24}
zepto	z	10^{-21}
atto	a	10^{-18}
femto	f	10^{-15}
pico	p	10^{-12}
nano	n	10^{-9}
micro	μ	10^{-6}
milli	m	10^{-3}
centi	c	10^{-2}
deci	d	10^{-1}
deca	da	10^{1}
hecto	h	10^{2}
kilo	k	10^{3}
mega	M	10^{6}
giga	G	10^{9}
tera	T	10^{12}
peta	P	10^{15}
exa	E	10^{18}
zetta	Z	10^{21}
yotta	Y	10^{24}

Appendix E.   Table of Laplace Transforms and Operations

The Laplace transform of f(t) is given by,

$$F(s) = \int_{-\infty}^{+\infty} f(t).\exp(-st)dt$$

and exists in the half-plane of the complex Laplace variable s for which the real part of s is greater than some fixed value.

f(t), t>0	Description	F(s)
Transforms:		
$\delta(t)$	Unit impulse at $t=0$	1
$\mu(t)$	Unit step at $t=0$	$1/s$
$\exp(-at)$	Signal decay	$1/(s + a)$
$1 - \exp(-at)$	Signal rise	$s/(s + a)$
t^n, $n \in Z^+$	Polynomial term	$n!/(s^{n+1})$
$\sin(\omega t)$	Sinusoidal waveform	$\omega/(s^2 + \omega^2)$
$\cos(\omega t)$	Cosinusoidal waveform	$s/(s^2 + \omega^2)$
$\sin(\omega t)/\omega$	-	$1/(s^2 + \omega^2)$
$\sinh(at)$	Hyperbolic cosine	$\omega/(s^2 - a^2)$
$\cosh(at)$	Hyperbolic sine	$s/(s^2 - a^2)$
$\sinh(at)/a$	-	$1/(s^2 - a^2)$
$t \exp(-at)$	-	$1/(s + a)^2$
Operations:		
$A f(t) + B g(t)$	Linear transform	$A F(s) + B G(s)$
$f'(t)$	First derivative	$s F(s) - f(+0)$
$f^{(n)}(t)$	nth derivative	$s^n F(s) - s^{n-1}f(+0) - ... f^{(n-1)}(+0)$
$\exp(at)f(t)$	-	$F(s - a)$

Appendix F.   Unit Conversion Factors

Unit	SI or cgs	fps
Length	1 cm = 0.39370 in	1 in = 2.5400 cm
	1 km = 0.62137 mi	1 mi = 1.6093 km
Area	$1 \text{ cm}^2 = 0.1550 \text{ in}^2$	$1 \text{ in}^2 = 6.4516 \text{ cm}^2$
	1 hectare = 2.4711 acre	$1 \text{ acre} = 4{,}046.9 \text{ m}^2$
Volume	$1 \text{ cm}^3 = 0.061024 \text{ in}^3$	$1 \text{ in}^3 = 16.387 \text{ cm}^3$
	1 litre = 0.21997 UK gallon	1 UK gallon = 4.5461 litre
		1 US gallon = 0.8327 UK gallon
Velocity	1 m/s = 2.2369 mile/hr	1 mile/hr = 0.44704 m/s
	1 km/hr = 0.62137 mile/hr	1 mile/hr = 1.6093 km/hr
Mass	1 kg = 2.20462 lb	1 lb = 0.45359 kg
	1 tonne = 0.98421 ton	1 ton = 1.0160 tonne
Density	$1 \text{ g/m}^3 = 0.036127 \text{ lb/in}^3$	$1 \text{ lb/in}^3 = 27.680 \text{ g/cm}^3$
Force	1 N = 0.22481 lb force	1 lbf = 4.4482 N
	$1 \text{ N} = 10^5 \text{ dyne}$	$1 \text{ dyne} = 10^{-5} \text{ N}$
	1 N = 7.2330 poundal	1 poundal = 0.13826 N
	1 N = 0.10197 kgf	-
Torque	1 N m = 0.7375 lbf ft	1 lbf ft = 1.356 N m
Pressure	$1 \text{ N/m}^2 = 1.4504 \times 10^{-4} \text{ lb/in}^2$	$1 \text{ lb/in}^2 = 6{,}894.8 \text{ N/m}^2$
	$1 \text{ kg/cm}^2 = 14.223 \text{ lb/in}^2$	$1 \text{ lb/in}^2 = 0.070307 \text{ kg/cm}^2$
	1 Pa = 0.0075006 torr	1 torr = 133.322 Pa
	$1 \text{ N/m}^2 = 10 \text{ dynes/cm}^2 = 1 \text{ Pa}$	-
	$1 \text{ Pa} = 10^{-5} \text{ bar} = 9.8692 \times 10^{-6} \text{ atmos}$	-
Energy	1 J = 0.23885 calorie	1 calorie = 4.1868 J
	$1 \text{ J} = 9.4781 \times 10^{-4} \text{ btu}$	1 btu = 1,055.1 J
	$1 \text{ J} = 10^7 \text{ erg}$	$1 \text{ erg} = 10^{-7} \text{ J}$
	1 kW h = 3.6 MJ	-
	1 J = 1 N m = 1 W s	-
	$1 \text{ eV} = 1.6021 \times 10^{-19} \text{ J}$	-
Viscosity	1 Pa s = 10 g/cm/s = 10 poise	$1 \text{ lbf s /in}^2 = 6{,}895 \text{ Pa s}$
Photometric	1 cd = 0.982 int. candles	1 candle = 1.018 cd
	$1 \text{ lx} = 0.09290 \text{ ln/ft}^2 \text{ or fc}$	1 fc = 10.764 lx
	1 cd = 1 lm/sr	-
	$1 \text{ lx} = 1 \text{ lm/m}^2$	-

Appendix G.   Fundamental Constants

Here is a list of the fundamental constants together with their value in SI units. Constants which are not explicitly used in the main text are denoted by putting the standard symbol in brackets. Basic properties of materials can be found in Tables 4.4 to 4.6 (§4.5).

Constant	Symbol	Value in SI units
Avogadro's number	(N_A)	6.0225×10^{23} per mole
Acceleration due to gravity	g	9.8067 m s^{-2}
Bohr magneton	(β)	1.165×10^{-29} Wb m
Boltzmann's constant	k	1.3805×10^{-23} J K^{-1}
Compton wavelength of electron	(λ_c)	2.4263×10^{-12} m
Electronic charge	e	1.6021×10^{-19} C
Electron charge-to-mass ratio	e/m_e	1.7588×10^{11} C kg^{-1}
Electronic rest mass	m_e	9.1096×10^{-31} kg
Electronic radius (classical)	(r_e)	2.8179×10^{-15} m
Electron volt	eV	1.6021×10^{-19} J
Faraday's constant	(F)	9.6487×10^4 C mol^{-1}
Fine structure constant	$\dfrac{e^2}{2h\varepsilon_0 v_c}$	7.2974×10^{-3}
Gas constant	(R)	8.3143 J K^{-1} mol^{-1}
Gravitational constant	(G)	6.670×10^{-11} N m^2 kg^{-2}
Impedance of free space	Z_0	376.73 Ω
Loschmidt's constant	(n_L)	2.6872×10^{-25} m^{-3}
Neutron rest mass	m_N	1.6748×10^{-27} kg
Permeability of free space	μ_0^B	$4\pi \times 10^{-7}$ H m^{-1}
Permittivity of free space	ε_0	8.8542×10^{-12} F m^{-1}
Planck's constant	h	6.6256×10^{-34} J s
Proton rest mass	m_P	1.6725×10^{-27} kg
Quantum charge ratio	h/e	4.1357×10^{-15} J s C^{-1}
Speed of light in vacuo	v_c	2.9979×10^8 m s^{-1}
Stefan-Boltzmann constant	(σ)	5.6697×10^{-8} W m^{-2} K^{-4}
Volume of 1 mole of ideal gas at STP	-	22.4 litres

Index

INDEX 323

Gravimeter, *see* mass microsensor

H
Hall
  angle, 202
  coefficient, 203
  devices, 202
    integrated, 205
  MOSFET, 213
  voltage, 203
Heat
  capacity
    table of, 70
  flow equation, 80
HEL, 241
HMAD 280
Hooke's law, 151
Hopping mechanism
  metal oxide films, 229
Human senses, 1,4
Humidity microsensor
  capacitive, 235
Hydrazine, 47

I
IEEE 488 interface, 33
IMAD 280
Inductive
  displacement microsensor, 158
Infra-red
  commercial devices, 144
  microsensor, 138
Input transducer, 2
Intelligent
  electronic nose, 283
  sensor, 269
    definition of, 270
  vision system, 290
Integrated sensors, 269
Interfacing
  parallel, 33
  RS-232, 32
  RS-422, 32
  RS-423, 32
  serial, 31
Ion
  beam milling

  reactor, 63
  etching, 62
  implantation, 38, 41
IPA, 48
IR, *see* infra-red
ISFET, 237
Isotropic wet etching, 45

J
Josephson
  effect, 200
  junction, 217
Joule effect, 200
Junction (diffusion depth), 40

K
Kelvin effect, *see* Thomson effect
Knoop hardness
  table of, 154

L
Langmuir-Blodgett film, 70
Laplace transform,
  table of, 316
Laser
  eximer, 66
Laser micromachining, 65
LEL, 241
Leptons, 118
Life
  characteristic, 259
LIGA technique, 75
$LiNbO_3$
  physical properties, 71
Lithography
  optical, 42
    resolution, 43
Loading roughness, 257
Load-strength distribution, 256
Lorentz force, 202
LPCVD, 44
LVDT, 27

M
MAD, 279
  see also array sensor
MAGFET, 213